Coping with Caveats in Coalition Warfare

Gunnar Fermann

Coping with Caveats in Coalition Warfare

An Empirical Research Program

Gunnar Fermann
Norwegian University of Science and
Technology
Trondheim, Norway

ISBN 978-3-030-06451-8   ISBN 978-3-319-92519-6   (eBook)
https://doi.org/10.1007/978-3-319-92519-6

Cover image © Sueddeutsche Zeitung Photo/Alamy Stock Photo
Cover design: Emma J. Hardy

Printed on acid-free paper

This Palgrave Macmillan imprint is published by the registered company Springer
International Publishing AG part of Springer Nature
The registered company address is: Gewerbestrasse 11, 6330 Cham, Switzerland

ACKNOWLEDGEMENTS

"No man is an island entire of itself."
I have incurred several debts in bringing this book-project to fruition. The idea for the book was conceived some three-years ago in collaboration with my previous master and Ph.D. student Per Marius Frost-Nielsen who was writing his dissertation on the same topic. His military background and academic proficiency made him a perfect partner in research. I dedicate the book to the confident and productive tutor–student relationship.

I wrote the bulk of the book in the academic year 2016–17 during my stay at the SCANCOR-Weatherhead Partnership at the Weatherhead Center for International Affairs at Harvard University. Weatherhead was very good to me. In particular, I am grateful to Professor Frank Dobbin, head of the partnership, for offering stimulating facilities, great company, and excellent academic advice on my work at several seminars. My appreciation also extends to the coordinator of the partnership, Catherine Himmel Nehring, who was indispensable in facilitating my stay. In seminars, I also received crucial feedback from my peers at Weatherhead including Ivar Bleiklie, Ulla Eriksson-Zetterquist, Antoinette Hetzler, Dennis Jancsary, Seija Kulkki, Samuli Patala, Elena Raviola, Gro Refsum Sanden, Tiina Ritvala, and Yamin Xie. Honorable mentioning is due to Stig Jarle Hansen who kept me up to date on several aspects of World Politics from his base on Belfer Center, Harvard University.

I am grateful to Professor Laura Neack for her very useful advice on the original book proposal, which impacted several lines of reasoning as well as the design of the volume. Draft manuscripts to the book have been reviewed in the panel on "Coalition and Alliance Politics" at the 2017 International Studies Association Convention in Baltimore, in the panel on "Foreign Policy—Nordic Perspectives and Beyond" at the 2017 Nordic Political Science Association Congress in Odense, and in the panel on "The Politics of Multinational Military Operations" at the 2018 International Studies Association Convention in San Francisco. I am particularly grateful for the comments offered by Thomas Forsberg, Jesse C. Johnson, Stephen M. Saideman, and Anders Wivel.

Close to home, I have for several years enjoyed collaboration within The Research Group on Foreign Policy and Security, Department of Sociology and Political Science, Norwegian University of Science and Technology (NTNU). I appreciate the multiple exchanges and feedback from colleagues Torbjørn L. Knutsen, Jo Jakobsen, Espen Moe, and Jonathon W. Moses on matters related to the topic of the book and beyond. Appreciated is also inputs and support from Gjert Lage Dyndal and Dag Henriksen of The Norwegian Armed Forces. I am grateful for the financial and administrative support to my 2016–17 sabbatical granted by The Department of Sociology and Political Science and The Faculty of Social and Educational Sciences, NTNU.

No man can do without a carrot, a mirage of what green pastures to walk after doing the hard work. Carrots came to me in different shapes. One is the former Marines, mechanic, and self-taught philosopher Kenneth C. Rush of The Republic of Texas. He made my motorbike ready for transcontinental riding and taught me how to ride in the backcountry of Texas, New Mexico, and Utah during my five-week crossing of the United States through 28 states.

The second carrot is coming home. A special thanks to Kristin and Jon to keep up with my traveling. The purpose of climbing real and proverbial mountains is to return to base to share the experience. The book is a partial fulfillment of this promise.

Trondheim, Norway   Gunnar Fermann
April 2018

CONTENTS

Abbreviations

AP	Alliance politics hypotheses
BPM	Bureaucratic politics model
CENTCOM	United States Central Command
CFP	Comparative Foreign Policy
CP	Comparative Politics
DEFCON	United States Defense Condition
DGP	Domestic and governmental politics hypotheses
EU	European Union
FPA	Foreign Policy Analysis
IR	International Relations
ISAF	International Security Assistance Force (Afghanistan)
JMAD	Joint method of agreement and difference
MA	Method of agreement
MCV	Method of concomitant variation
MD	Method of difference
NATO	North Atlantic Treaty Organization
NORDBAT	Nordic battalion in UNPROFOR
OEF	Operation Enduring Freedom (Iraq)
OPM	Organizational process model
PI	Politics of implementation hypotheses
PMRoE	United States Worldwide Peacetime Rules of Engagement for Seaborn Forces
PRoE	United States Peacetime Rules of Engagement
QCA	Qualitative Comparative Analysis
RAM	Rational actor model
RoE	Rules of engagement

SFOR	Stabilization Force (Bosnia)
SOP	Standard operating procedure
SPM	Scope for political maneuvering
SRoE	United States Standing Rules of Engagement
UNEF	United Nations Emergency Force (Sinai)
UNFICYP	United Nations Force in Cyprus
UNIFIL	United Nations Interim Force in Lebanon
UNOSOM	United Nations Operation in Somalia
UNPROFOR	United Nations Protection Force (Balkans)
US	United States

LIST OF FIGURES

LIST OF TABLES

PART I

Introduction

CHAPTER 1

Making Sense of the Politics of Caveats

Pooling resources to deter external threats to security and to safeguard common interests is as old a phenomenon as organized warfare (Parker 2005). Still, alliance politics have always been pervaded by coordination problems because sovereign states interacting to realize some public good more often than not have both shared and diverging interests (Snidal 1985). Even within the same polity, inclinations may be ambiguous and interests only partially congruent. Difficult collaboration is very much part of the human condition.

The shared perception of some common threat is the principal glue of security alliances. Crucial is also the integrating capabilities of the alliance-leading hegemonic power, whether it leads through attraction, persuasion, or command. The centrifugal forces in alliances relate to the existence of conflicting, individual self-interests among alliance partners concerning what should be the purpose of the alliance, how to run it, and how costs, risks, and gains are distributed among alliance members in a particular coalition context. Depending on the robustness and flexibility of the security alliance and the coalition forces in question, such intra-alliance policy conflicts may reduce the credibility of the coalition, as well as its military efficiency and capacity to serve political goals. Over time, intra-alliance contradictions as expressed in less than successful coalition operations may undermine the relevance of the security arrangement.

The Prussian General, Carl von Clausewitz, had the first-hand experience from how alliance dilemmas at the collective and individual level

© The Author(s) 2019
G. Fermann, *Coping with Caveats in Coalition Warfare*,
https://doi.org/10.1007/978-3-319-92519-6_1

play out on the field against better organized, led, and motivated French forces. He survived to tell about it as military theorist:

> It would all be tidier, [...] if the contingent promised – ten, twenty, or thirty thousand men – were placed entirely at the ally's [the Force Commander's] disposal and he was free to use it as he wished. [...]. But that is far from what happens. The auxiliary force usually operates under its own [national] commander [who] is dependent only on his government and the objective the latter set him will be as ambiguous as of its aims. (Von Clausewitz 1976 [1832]: 250)

Clausewitz is drawing attention to a theater of war in which what is politically feasible to agree upon among alliance partners, does not allow for the full use of pooled military resources for political purposes. Dual chains of command and only partially shared operational commitments create uncertainty and make it difficult for coalition Force Commander to utilize national contingents in meaningful ways. We see this deficit of what is politically feasible to agree upon played out also in contemporary Western burden-sharing debates, and in command—and control issues discussed in connection with the establishment and running of coalition forces (Driver 2016; Frost-Nielsen 2016, 2017).

Restrained participation in coalition operations come in two main shapes, short of nonparticipation. One is the provision of a limited number of personnel in noncombatant functions, such as military medics and staff officers. Miniscule and symbolic contributions may also take the shape of modest logistical or financial support from governments wanting to demonstrate some political support. For the risk-averse, lukewarm, or skeptical ally, another kind of restrained participation is available for consideration. The option is the attachment of conditions—"caveats"—as to when, where, and how its substantial military contingent be used by coalition Force Commander in the theater of war. This phenomenon is the focal point of the present study. For now, we shall proceed on the conception that *caveats are national reservations on the use of force in a coalition context* (Frost-Nielsen 2016: 14–16; 2017: 3–4). Several topical research questions come to mind: How may the application of caveats on national military contingents affect coalitions? How widespread is the use of national reservations on the use of force? How may Force Commander and political decision-makers manage and compensate for the negative impact of caveats, if at all? What conditions governments' use of caveats in coalition operations? How may we go about to better understand the politics of caveats?

The latter question is about how to approach and design research on the politics of caveats, and is the main concern of the present study.

THE INCONVENIENCE OF CAVEATS IN COALITION OPERATIONS

Since the end of the Cold War, NATO, the United Nations and ad hoc "Coalitions-of-the-Willing" have engaged forcefully in conflicts in Kosovo, Afghanistan, Iraq, and Libya, to mention but some of the most broadcasted interventions. Besides the fundamental problem of enforcing political solutions in countries pervaded by ethnic cleavages, economic underdevelopment, corruption, and weak political institutions, the multitude of national restrictions on the use of force applied by coalition partners in International Security Assistance Force (ISAF), Afghanistan 2001–2014, contributed to the less than successful implementation of the mission's political goals (Auerswald and Saideman 2014).

There are reasons to believe that the use of caveats in coalition operations have been significant in the post-Cold War era. At one point, NATO generals in Afghanistan put together an eighty-page document describing 70 instances of national reservations on the use of force in the coalition operations in this particular theater of war (Bergen 2011: 189). David P. Auerswald and Stephen M. Saideman (2014) found on their count some 50 caveats applied in the coalition operations in Afghanistan, and Otto Trønnes (2012) discovered tenfold of instances of caveats in Norway's behavior as a coalition partner in various multinational military operations. In a comparative study of caveats in the coalition operations in Libya 2011, Per Marius Frost-Nielsen (2017) found the Netherlands to apply heavy restrictions on the use of airpower. Germany used restrictive caveats on how their forces could be used to limit domestic political risks at the expense of the operational effectiveness of the coalition forces in Afghanistan (Lombardi 2008).

Among coalition Force Commanders, caveats are seen mainly as a severe impediment to military flexibility and efficiency, and thus to the successful implementation of the political mandate of coalition forces. Referring to coalition-force Operation Enduring Freedom (OEF) in Iraq 2003–2010, US Secretary of Defense Donald Rumsfeld pointed to the use of national caveats as "a quite complex problem for the [force] commander" to cope with (US DoD 2005). The question of caveats again appeared high on the agenda of the NATO Summit in Riga in 2006

(NATO 2006), thus confirming caveats as a continuous challenge to the effective use of coalition forces (Clark 2001). Furthermore, the imposition of national reservations on the use of force in the Stabilization Force (SFOR) in Bosnia, in 1990s, was less diplomatically described as "cancer that eats away at the effective usability of troops" (Johnson 2004) by limiting "the tactical commanders' operational flexibility" (Jones 2004).

The UN operation in Somalia in the early 1990s (UNOSOM) may serve to illustrate the point. In this operation, several governments frequently intervened in the UN chain of command to make sure their military contingents were kept out of harm's way. In 1993, the UN took action and dismissed the Italian Force Commander because of his failure to obey direct orders from UN headquarters in New York (Von Hippel 2000: 75). Research indicates that the Italian case of insubordination—a matter of split loyalties—was not exceptional. Similar instances of deviant practicing of coalition's common rules of engagement (RoE) have been revealed in several other UN operations (Feldman 2008; Chopra et al. 1995: 72; Findlay 2002: 178, 183; Hirsch and Oakley 1995: 63–66, 75–76, 82–83), including United Nations Interim Force (UNIFIL) in Lebanon, and United Nations Protection Force (UNPROFOR) in the Balkans (Findlay 2002: 117–118, 134–135). Furthermore, during NATO's 1999 air operations against Serbia in the conflict over Kosovo, some governments did not allow for their aircraft or bases to bomb particular targets. Other governments would not allow own aircraft to attack specific targets but accepted that other states used aircraft for offensive purposes operating from the same base (Lambeth 2001: 185–189; Weitsman 2014: 74–83).

In particular, the extensive use of caveats in Afghanistan represented a serious challenge to the military effectiveness of coalition forces, and created political tensions between coalition partners concerning the sharing of burdens and risks. Due to the extent of caveats applied, and the long-lasting military commitment in Afghanistan, the practice of caveats became much more visible in ISAF compared to previous coalition operations. The high visibility of the phenomenon of national reservations on the use of force did much to increase the awareness of military and political decision-makers who increasingly came to debate the phenomenon in terms of "caveats" or "national caveats" (Hoehn and Harting 2010: 53–55; Morelli and Belkin 2009: 10–12; Ringsmose 2010). Analysts within the field of intra-alliance burden-sharing picked upon this vocabulary, and used it to emphasize the problems within NATO to cooperate effectively in Afghanistan (De Borchgrave 2009; Hunter 2008; Joyner 2009; Noetzel and Scheipers 2007).

In 1990–1991, US-led coalition intervention in Iraq, the United States instigated the mapping of how members of the coalition understood the common RoE. This measure was invited to minimize coordination problems and difficulties in the implementation of the military operations (Humphries 1992; Phillips 1993: 24; Weitsman 2014: 50–55). Knowing the positions of member states on the interpretation of common RoE, made it possible for Force Commander to integrate national military contingents in ways that accounted for what he could expect the different contingents to be willing to do. National contingents with the more restrictive reservations on the use of force were deployed in areas of low military risk and assigned less demanding tasks. In this way, hesitant military participation, although politically valuable, would not hurt the operational flexibility of the coalition force more than necessary (Bennett and MacDonald 1995: 125).

Per Marius Frost-Nielsen notes that to bridge diverging political concerns, the mandates of several UN operations through the 1990s were intentionally ambiguous (2016: 5–6). It was left to coalition Force Commander to operationalize the overtly vague mandate into a feasible military operational concept, the common RoE of the force as a whole. However, even the most skilled translation of the mandate to a common RoE could not completely circumvent the political reality that some troop-contributing governments were less enthusiastic, and more vulnerable to critique from domestic constituencies than other governments. Add to this that national military organizations vary in strategic culture and legal traditions. Such factors contributed to divergence in national interpretations of the common RoE of the military operation (Soeters et al. 2010; Soeters and Manigart 2008; Zinni and Lorenz 2000: 223–244; Findlay 2002: 368–374; Lorenz 1995: 22; Palin 1995: 34–35; Dworken 1994: 34).

"National Roe" came to conceptualize the divergent national interpretations and practicing of the "common RoE" of UN peace operations. Lukewarm governments would interpret "common RoE" more restrictive than the more enthusiastic governments. From this followed burden-sharing conflicts, and coordination problems in the field, not qualitatively different from those later reported as "caveats" in the context of Afghanistan (Frost-Nielsen 2016: 5–6).

For long, the existence of different interpretations and military practicing of the common RoE was conveniently explained away as the result of the uncoordinated military action, or lack of operational skill and resources. If there were a need to de-politicize, the language used did serve the purpose. This narrative was in marked contrast to the politicized language used toward member states in ISAF, which applied restrictive caveats on their use of force as compared to the common RoE of this coalition force. In Afghanistan, "caveats" were described within a burden-sharing narrative as a deliberate failure to take full collective responsibility due to self-serving national priorities.

Contrasting the UN operations of the 1990s with the NATO operations in Afghanistan in the 2000s, the following pattern stands out: Different agencies (UN vs. NATO/"Coalitions-of-the-Willing") suffer similar problems (reduced military efficiency and effectiveness) but apply divergent problem-ascribing narratives (functional vs. political), and different terms to denote the symptom of political disagreements within coalition operations ("national RoE" vs. "caveats").

Even if some of the national deviations in the interpretation of common RoE were due to lack of coordination and differences in military competence, much evidence indicates that also in UN operations restrictive practicing of common RoE be traced back to political calculations of governments. In the so-called "Brahimi-report," which was a comprehensive evaluation of the UN experiences with peace operations through the 1990s, the real problem of lukewarm participation was framed as a problem of national political interference in coalition chain of command:

> Troop contributors must ensure that the troops they provide fully understand the importance of an integrated chain of command, the operational control of the Secretary-General and the standard operating procedures and rules of engagement of the mission. It is essential that the chain of command in operation be understood and respected, and the onus is on national capitals to refrain from instructing their contingent commanders on operational matters. (UN 2000: 45)

By the year 2000, it was time to speak out about what for long had become increasingly clear: Not all deviations from the common RoE in UN operations were due to incompetence or lack of coordination. The "national RoE" practiced by governments contributing to UN peace operations in the 1990s were partly, if not predominantly, politically motivated. The ingredient of political calculation makes the extensive use of "national RoE" in UN peace operations functionally equivalent, if not identical, to the "caveats" applied by governments in coalition operations in Afghanistan, Iraq, and Libya. Politically motivated deviations from common RoE are expressions of national reservations on the use of force whether the phenomenon observed in a UN context, or in coalition contexts, such as in Afghanistan, Iraq, Libya, and Syria.

The phenomenon of national reservations on the use of force in coalition contexts is traceable deep down in history. Among the states that made up the coalition against Napoleon in the early nineteenth century, it was not unheard of that a member of the alliance in the midst of battle negated the supreme commander's tactical dispositions (Riley 2007). Similar instances of reluctance to subordinate national contingents to a joint command led by a foreign military commander is found among the Éntente powers France and Britain during the First World War. Subordination of national forces under a single unified military command to increase military efficiency was difficult to achieve because the allies were preoccupied with securing national influence on strategic and tactical military decisions (Bliss 1922). The result was a suboptimal arrangement, where national officers represented in the coalition chain of command and assigned veto powers on the use of their forces. Finally, despite the close military relationship between the United States and Britain during the Second World War, the coalition partners disagreed at several junctions on who should coordinate and lead the war effort in various theaters of war (Kennedy 1983: 11–14; Stoler 2000). This state of affairs is likely to have triggered the use of caveats at multiple crossroads.

The parallel to Carl von Clausewitz' narrative on early nineteenth century alliance politics could not be more evident. Similar observations across centuries indicate that national reservations on the use of force within allied military operations are as old a phenomenon as alliances and coalition operations themselves. However, we have reason to suspect that there is a variation in the use of caveats across states, coalition forces, and over time. We would like to know how and why. These are empirical puzzles worthwhile for systematic study. Not least because coalition operations are the dominant and most broadcasted form of military power-projection internationally. Undoubtedly, the political study of caveats can make use of insights from research on alliance politics, civil–military relations, and constitutional and international regulation of the use of force, to mention but a few scholarly neighborhoods. Prior to engaging adjacent bodies of literature, we need to know what the literature on the politics of caveats and coalition participation itself have to offer, so fare.

RESEARCH ON CAVEATS: CATCHING UP WITH THE PAST AND THE PRESENT

Research on phenomena resembling national reservations on the use of force in coalition operations is closely related to post-Cold War operations within the framework of the UN, NATO, or some "Coalition-of-the-Willing" context. Due to the significant operational and political challenges national reservations on the use of force represents to military coalitions, caveats been frequently mentioned in the broader security literature since the early 2000s. References to caveats are found in research on civil–military relations in complex military operations (Ruffa et al. 2013); counterinsurgency doctrine and state-building activities (Meyer 2013); combat effectiveness and multinational operations (Marten 2007; Deni 2004); multilateral military intervention and burden sharing (Richter and Webb 2014; Kreps 2008); and within the comparative study of democratic participation in armed conflict (Mello 2014).

Since the early 2000s, it is not hard to find research that deals with caveats in particular. NATO operations in Afghanistan from 2003 have attracted considerable research in this regard. Several scholars took note of the sharp rise in the application of national reservations on the use of force, and made it a part of their research agenda (Findlay 2002; Auerswald 2004; Bird 2007; Nevers 2007; Kreps 2008; Ringsmose and Thruelsen 2010; Williams 2011; Rynning 2012; Saideman and Auerswald 2012; Dorman 2012; Knudsen and Klingenberg 2013; Weitsman 2014; Auerswald and Saideman 2014).

However, with some notable exceptions such as Auerswald and Saideman (2014), Saideman and Auerswald (2012), Findlay (2002), and Frost-Nielsen (2011, 2013, 2016, 2017), Political Science scholars have yet to engage forcefully in the systematic and focused study of caveats. We do not know nearly enough about the politics of caveats, what political purposes caveats serve for foreign policy decision-makers, and what structural and institutional factors may condition the use of national reservations in statecraft. This state of affairs needs to be improved upon if we are to understand and advise on, not only the negative consequences of caveats on coalition forces, but also the trade-offs and possibilities for political engineering, related to the application of caveats in foreign policy and alliance politics.

The literature on caveats is more descriptive than explanatory, more focused on serving immediate political and operational needs than long term, systematic knowledge-building, and more preoccupied with the immediate detrimental effects of caveats on burden sharing, military efficiency, and political effectiveness than in unraveling the conditions for states' applying reservations on the use of force in the first place. In different ways, these limitations seem to relate to the fact that caveats often dealt with as an auxiliary phenomenon in the research literature.

A fundamental analytical limitation in research bordering on national reservations on the use of force is the inconsistent use of the key concept of "caveats." It is thus not clear to what particular aspect of the phenomenon the term refers. Frost-Nielsen (2016: 4–5) finds that "caveats" is applied to point out how national reservations prevented national contingents participate in offensive and risky military operations (Mello 2014: 113–114; Ringsmose 2010: 328; Sky 2007: 16), and that others have used the concept of "caveats" concerning how states place national officers in the coalition chain of command to execute some discretion as to how Force Commander may use their military contingents (Høiback 2009: 23–34; Saideman and Auerswald 2012: 69–79; Young 2003: 115). Some scholars relate caveats to constitutional limitations of the use of force in states, such as in Germany (Koschut 2014: 351–354) and Japan (Van der Meulen and Kawano 2008). One contribution mentions "caveats" in the context of financial and logistical limitations on the military contribution. However, without discussing whether politically imposed reservations on the use of force may belong to another class of phenomenon than economic constraints and logistical deficiencies (Brophy and Fisera 2010: 1). Finally, in Afghanistan, the perhaps most frequently used conception of "caveats" relates to the limitations a contributing state may put on where its forces are allowed to operate (Kay 2013: 109–110; Noetzel and Rid 2009: 75; Noetzel and Schreer 2009: 532).

Hence, "caveats" is an interesting, complex and—for the time being—conceptually confused phenomenon. The concept of caveats lacks settled boundaries. We cannot establish the politics of caveats as a research domain in its own right as long the subject of study not precisely identified. When still lacking agreement on the definition and operationalization of the key concept in the research field, it is premature to start building systematic databases on the frequency and variation of the phenomenon, let alone reason what explanatory approaches and methodological designs might give more epistemological traction.

The key to rectifying conceptual confusion is to study the relevant bodies of literature, juxtapose divergent suggestions against the empirical record, and come up with a reasoned conception of caveats that is precise enough to distinguish caveats from adjacent phenomena, and inclusive enough to discriminate between different subclasses of caveats which might have different causal origines and impacts.

This brief review suffices to indicate that the inclination to politically control national contributions to coalition operations is a generic phenomenon across otherwise different contexts where stakes are extremely high. There are ways of minimizing the detrimental consequences of this disintegrating impulse. The strength of the impulse may vary depending on circumstances, not the least on the nature of the common threat faced. However, the fact remains that sovereign states rarely have completely overlapping interests in alliance and coalition politics. In this perspective, the foreign policy instrument of caveats may turn out to become a blessing in disguise for the coalition in that restrictive caveats allow the coalition member to fine-tune its contribution to optimize the balance between different concerns and interests. If caveats were not an option, we might expect more governments to defect from the coalition, or only make a symbolic contribution to the war effort. This line of reasoning comes out as a defense for the foreign policy instrument of caveats in coalition operations and is an important take-home message from the study.

THE AIM OF THE STUDY: TOWARD AN EMPIRICAL RESEARCH PROGRAM

The aim of the present study is to offer reasoned answers to the need for systematic research on the politics of caveats—conceptually, analytically, theoretically, and methodologically. We have indicated that the political study of caveats is a rather nascent field of research. There are some excellent studies to relate to and more if we count studies treating caveats as an afterthought, an auxiliary phenomenon, or a case-specific peculiarity. Still, the field is fragmented, and we see too little of the conceptual, theoretical and epistemological self-reflection characterizing more mature fields of research.

Scholarly debates should include vivid discussions on the conceptualization of the phenomenon under study, and on how key variables be operationalized for empirical research. We may consider caveats in terms of dependent or independent variable. We should attempt to develop a limited number of generic research questions that particular case-study can relate to. We probably need to reason more about what knowledge gained by employing theories of Comparative Politics (CP), the systemic approach of International Relations (IR), and approaches taking the agency of the state as the main point of departure for the study of caveats. Research-wise, we must deal with the fact that states and international security organizations consider information on caveats militarily and politically sensitive. How can we make sure that the most interesting data is accessible to us? Without some clever data-gathering strategies, any approach to the study of caveats come up empty-handed. As to the empirical analyses of caveats-relevant data, we should inspire discussion as to what methods for causal inference are likely to offer the better epistemological traction—given the present state of caveats research and the framework for analysis applied.

On this backdrop, the study of the politics of caveats would seem to be fertile ground for a scholarly contribution with a programmatic purpose, arguing directions for future research. In the present study, we offer a proposal for *an empirical research program on the foreign politics of caveats*. An "empirical research program" describes a problem-solving research approach directed toward shedding light on a phenomenon or a limited set of phenomena (Lakatos 1978). This effort includes, first, the contextualization and conceptualization of the phenomenon under scrutiny. Second, the working out of a framework for analysis, including a general approach and several bodies of middle-range theory from which empirical propositions deduced on conditions for the application of caveats in statecraft. Finally, the indication of promising research strategies and methods for the gathering and analyses of data that speaks to the nature of the phenomenon under scrutiny, takes into consideration the methodological implications of the analytical framework chosen, and account for the fact that the political study of caveats is in an early stage of cumulative research.

We argue that the bottom-up approach of Foreign Policy Analysis (FPA) is well equipped as a stepping stone for the systematic study of the politics of caveats: FPA ascribes agency to the foreign policy-making state and treats caveats as an instrument of foreign policy. FPA accounts for both domestic and global political factors in explaining foreign policy-making. The approach is preoccupied with tracing the influence of exogenous factors from both the global and domestic environments that influence the decision-making and implementing processes of the foreign policy-making state. As we will elaborate on in Part Three of the study, the multi-level and decision-making approach of FPA is extremely eclectic and ambitious in theoretical scope. The FPA approach is also thorough in the empirical tracing of decision-making processes influenced by global politics, domestic politics, and institutional inertia at the level and different phases of decision-making and implementation (Fermann 2010, 2013: 89–140; Neack 2013; Carlsnaes 2002, 2008; Hudson 2005, 2007; Hill 2003; Webber and Smith 2002; Kubálková 2001; Clarke and White 1989).

In the context of the politics of caveats, FPA inspires one research question in particular: *What political goals and functions do caveats serve for foreign policy decision-makers in balancing external and domestic concerns as perceived through the lenses of national and institutional interests?* Put differently: What decision-making problems and opportunities may the application of national reservations on the use of force (caveats) address for foreign policy decision-makers, who consider contributing a military contingent to some multinational military operation?

In his Ph.D. dissertation, Per Marius Frost-Nielsen studied this puzzle through a string of single and comparative case studies. Three initial speculations survived our multiple discussions (2016: 6):

- May caveats limit the political costs of participating in coalition forces motivated by alliance obligations?
- May caveats facilitate the construction of domestic winning political coalitions required to participate in allied coalition operations?
- May caveats secure some national control of contingents put under a common military chain of command?

Indeed, the overarching idea of the present study is that caveats are a foreign policy instrument used by strategic decision-makers at the national level to balance different concerns in alliance politics, domestic politics, and in the politics of controlling the sword in the theater of war.

Most of the academic research on the politics of caveats draws upon insights from IR and CP approaches. In the context of caveats, the approach of FPA represents a somewhat novel research-angle, which arguably is more realistic than the systemic IR approach because FPA since its inception has gravitated toward traveling several levels of analyses, including the level of governmental politics where a lot of caveats-relevant decision-making is going on. It is notable that the IR community in the late 1980s celebrated Robert D. Putnam's "two-level game" approach to the study of international negotiations as an analytical breakthrough (Putnam 1988), without simultaneously acknowledging that cross-level analysis has been bread and butter in FPA since the late 1950s, although with different explanandum in mind.

In contrast to the "top-down" approach of IR, where caveats theorized primarily as a phenomenon making international collaboration and the implementation of mandates more difficult, we argue that the "bottom-up" approach of FPA is tailor-made to shed light on caveats as a problem-solving instrument within the limits of what the foreign policy-making state deems politically feasible.

While CP already has brought valuable contributions to the emerging field (e.g., Auerswald and Saideman 2014), FPA is better equipped to unravel the causal mechanisms at work in-between cause and effect in the politics of caveats (Fermann 2013). In CP, such mechanisms are theoretically argued but rarely empirically traced and confirmed in the empirical study of policy-making processes. Indeed, FPA is designed to come to grips with real-life policy-making and implementing processes utilizing process-tracing case-study designs.

At its most thorough, the empirical tracing of decision-making and implementing processes may run as follows: How does this and that structural and motivational factor in the global and domestic environments of the decision-making process, influence this and that decision maker or decision-making institution, in what phase of the decision-making and implementing processes, through what mechanisms, to impact the formulation of preferences and the choice of policy instruments, and to what effect on foreign policy actions? FPA is capable of detecting inconsistencies in the relationships between policy goals, means, and implementation, and considers it an empirical question whether such discrepancies are due to political mismanagement, political intention, or the fog of war.

In a still nascent research field such as ours, it is an unfortunate omission that we see few multi-level and process-tracing studies in the politics of caveats (Frost-Nielsen 2011, 2013; Husby 2015). We need such studies to capture the interplay between domestic and global politics on the decision-making process mostly, but not entirely neglected in IR and CP. We need process trace decision-making processes to reality check the mechanisms theoretically assumed in generically ambitious CP studies. However, we do not claim that the choice of approach to caveats is a question of one or the other. Rather, it is clearly a question of complementarity. Nevertheless, an empirical research program built on the ontological foundation of the FPA approach may attract new scholars to researching the politics of caveats, and produce novel insights from angles not properly covered by IR and CP approaches.

In particular, FPA frames caveats as an instrument of political engineering that allows foreign policy-makers better to cope with difficult decision-making challenges. The complicated and sometimes critical nature of foreign politics is because foreign policy-making is pinched between a rock and a hard place, between domestic and global politics. In alliance politics, the stakes can be extremely high. In particular, in questions relating to whether and how a sovereign state is to contribute lethal force within the framework of coalition operations. In domestic politics, stakes may be high as well. Publicized collateral damage and loss of own military personnel may threaten the continued existence of the responsible government in question.

OUTLINE

The essence of an empirical research program is to provide directions and suggestions on how fruitfully study some social phenomenon. The empirical research program on the politics of caveats does not by itself pretend to provide conclusive answers to substantial research questions. It is mainly for subsequent research to deliver empirical analyses of substantial research questions. The contribution of the empirical research program is mainly to be assessed on its capacity to inspire, direct, reason, and facilitate systematic research on the politics of national reservations on the use of force in coalition operations.

The several lines of argument making up the empirical research program run through 10 chapters between the introductory and concluding chapters. The structure of the study reflects the four main epistemological steps of research program—conceptual delimitation of the phenomenon of caveats, reasoning and explanation of the foundational approach of FPA, theoretical arguing of empirical propositions, and the working out of promising methodological strategies for empirical research. For readers inclined to jump to conclusions in the very literal sense, Chapter 12 offers an economical recapitulation of the programmatic contributions.

In Chapters 2–4 (Part 2: Conceptualizing caveats), the analytical context of caveats is discussed and the concept of caveats defined and operationalized.

In Chapters 5–7 (Part 3: Approaching caveats), we reason the approach of FPA in terms of the non-refutable "hard core" of the empirical research program, discuss the analytical components in the foreign policy-making and implementing processes, and argue the translation of Political Science middle-range theory from several levels of analyses to properly explain foreign policy-making processes and outcomes.

In Chapters 8–10 (Part 4: Theorizing caveats), selected bodies of theory are applied to the politics of caveats. Three sets of hypotheses are generated on the assumption that caveats may limit the foreign policy-making state's political costs of participating in coalition forces motivated by alliance obligations; facilitate the construction of domestic winning political coalitions required to participate in allied coalition operations; and secure some national control of contingents put under a common military chain of command.

In Chapter 11 (Part 5: Researching caveats), we review, explain, and recommend methods for the gathering and analyses of data that takes into account the approach applied and the state of the research field in question.

REFERENCES

Auerswald, D. P. (2004). Explaining Wars of Choice: An Integrated Decision Model of NATO Policy in Kosovo. *International Studies Quarterly, 48*(3), 631–662.

Auerswald, D. P., & Saideman, S. M. (2014). *NATO in Afghanistan: Fighting Together, Fighting Alone*. Princeton, NJ: Princeton University Press.

Bennett, D. A., & MacDonald, A. F. (1995). Coalition Rules of Engagement. *Joint Force Quarterly*, Summer(8), 124–125.

Bergen, P. L. (2011). *The Longest War*. New York, NY: Free Press.

Bird, T. (2007). The European Union and Counter-Insurgency: Capability, Credibility, and Political Will. *Contemporary Security Policy, 28*(1), 182–196.

Bliss, T. H. (1922). The Evolution of the Unified Command. *Foreign Affairs, 1*, 1–30.

Brophy, J., & Fisera, M. (2010). *National Caveats and Its Impact on the Army of the Czech Republic*. http://user.unob.cz/fisera/files/clanky/National_Caveats_Short_Version_version_V_29JULY.pdf.

Carlsnaes, W. (2002). Foreign Policy. In W. Carlsnaes, T. Risse, & B. A. Simmons (Eds.), *Handbook of International Relations*. London: Sage.

Carlsnaes, W. (2008). Actors, Structures, and Foreign Policy Analysis. In S. Smith, A. Hadfield, & T. Dunne (Eds.), *Foreign Policy: Theories, Actors, Cases* (pp. 83–100). Oxford: Oxford University Press.

Chopra, J., Eknes, Å., & Nordbø, T. (1995). *Peacekeeping and Multinational Operations*. Oslo: Norwegian Institute of Foreign Affairs.

Clark, W. K. (2001). *Waging War: Bosnia, Kosovo, and the Future of Combat*. New York, NY: Public Affairs.

Clarke, M., & White, B. (1989). *Understanding Foreign Policy: The Foreign Policy Systems Approach*. Aldershot: Edward Elgar.

De Borchgrave, A. (2009). *Commentary: NATO Caveats*. UPI. Retrieved from http://www.upi.com/Top_News/Analysis/de-Borchgrave/2009/07/10/CommentaryNATO-caveats/47311247244125/.

De Nevers, R. (2007). NATO's International Role in the Terrorist Era. *International Security, 31*(4), 34–66.

Deni, J. R. (2004). The NATO Rapid Deployment Corps: Alliance Doctrine and Force Structure. *Contemporary Security Policy, 25*(3), 498–523.

Dorman, A. M. (2012). NATO's 2012 Chicago Summit. *International Affairs, 88*(2), 301–312.

Driver, D. (2016). Burden Sharing and the Future of NATO: Wandering Between Two Worlds. *Defense & Security Analysis, 32*(1), 4–18.

Dworken, J. T. (1994). Rules of Engagement—Lessons from Restore Hope. *Military Review, 74*(September), 26–34.

Feldman, R. L. (2008). Problems Plaguing the African Union Peacekeeping Forces. *Defense and Security Analysis, 24*(3), 267–279.

Fermann, G. (2010). Strategisk ledelse i utenrikspolitisk perspektiv. I Gjert Lage Dyndal (Ed.), *Strategisk ledelse i krise og krig* (pp. 9–61). Bergen: Fagbokforlaget.

Fermann, G. (Ed.). (2013). *Utenrikspolitikk og norsk krisehåndtering*. Oslo: Cappelen Damm Akademika.

Findlay, T. (2002). *The Use of Force in UN Peace Operations*. Oxford: Oxford University Press.

Frost-Nielsen, P. M. (2011). Politisk kontroll av militær deltakelse i internasjonale operasjoner: Restriksjoner på bruk av norske kampfly i Afghanistan. *Internasjonal Politikk, 69*(3), 359–386.

Frost-Nielsen, P. M. (2013). Norske kampfly i Afghanistan 2006. In G. Fermann (Ed.), *Utenrikspolitikk og norsk krisehåndtering* (pp. 267–298). Oslo: Cappelen Damm Akademika.

Frost-Nielsen, P. M. (2016). *Betingede forpliktelser. Nasjonale reservasjoner i militære koalisjonsoperasjoner*. Ph.D. Dissertation in Political Science, Department of Sociology and Political Science, Norwegian University of Science and Technology (NTNU), Trondheim.

Frost-Nielsen, P. M. (2017). Conditional Commitments: Why States Use Caveats to Reserve Their Efforts in Military Coalition Operations. *Contemporary Security Policy, 38*(3), 371–397.

Hill, C. (2003). *The Changing Politics of Foreign Policy.* London: Palgrave Macmillan.

Hirsch, J. L., & Oakley, R. B. (1995). *Somalia and Operation Restore Hope— Reflection on Peacemaking and Peacekeeping.* Washington: United States Institute of Peace.

Hoehn, A. R., & Harting, S. (2010). *Risking NATO—Testing the Limits of the Alliance in Afghanistan.* Santa Monica: RAND Corporation. https://www. rand.org/content/dam/rand/pubs/monographs/2010/RAND_MG974.pdf.

Hudson, V. M. (2005). Foreign Policy Analysis. Actor-Specific Theory and the Ground of International Relations. *Foreign Policy Analysis, 1*(1), 1–30.

Hudson, V. M. (2007). *Foreign Policy Analysis: Classical and Contemporary Theory.* Boulder, CO: Rowman and Littlefield.

Humphries, J. G. (1992). Operations Laws and the Rules of Engagement in Operations Desert Shield and Desert Storm. *Airpower Journal, 11*(3), 25–41.

Hunter, R. E. (2008). NATO Caveats Can Be Made to Work Better for the Alliance. *European Affairs, 9*(1–2). https://www.europeaninstitute.org/ index.php/42-european-affairs/winterspring-2008/68-nato-caveats-can-be-made-to-work-better-for-the-alliance.

Husby, G. (2015). *Fra hull I luften, til hull I Gaddafis bunker. Bruk av politiske reservasjoner på norsk militærmakt I flernasjonale koalisjonsoperasjoner. En komparativ studie av F-16 bidragene i Kosovo, Afghanistan og Libya.* Master Thesis in Political Science, Department of Sociology and Political Science. Norwegian University of Science and Technology (NTNU), Trondheim.

Høiback, H. (2009). The Noble Art of Constructive Ambiguity. *Oslo Files on Defence and Security, 3,* 19–39.

Johnson, G. G. (2004). Examining the SFOR Experience. NATO. http://www. nato.int/docu/review/2004/Historic-Changes-Balkans/Examining-SFOR-experience/EN/index.htm.

Jones, J. L. (2004). Prague to Istanbul: Ambition Versus Reality. Global Security: A Broader Concept for the 21st Century. Center for Strategic Decision Research 21st International Workshop on Global Security—Berlin, 7–10 May. http://csdr.org/2004book/Gen_Jones.htm.

Joyner, J. (2009). *Afghanistan Caveats Coming to End.* Atlantic Council. 10 July. http://www.atlanticcouncil.org/blogs/new-atlanticist/afghanistan-caveats-coming-to-end.

Kay, S. (2013). No More Free-Riding: The Political Economy of Military Power and the Transatlantic Relationship. In J. H. Matlary & M. Petersson (Eds.), *NATO's European Allies—Military Capability and Political Will* (pp. 97–120). Hampshire: Palgrave Macmillan.

Kennedy, P. (1983). Military Coalitions and Coalition Warfare Over the Past Century. In K. Neilson & R. A. Prete (Eds.), *Coalition Warfare—An Uneasy Accord* (pp. 1–15). Waterloo: Wilfrid Laurier University Press.

Knudsen, E., & Klingenberg, S. (2013). *Cooperating in War—Coalition Warfare in Afghanistan*. Copenhagen: Forsvarsakademiet.

Koschut, S. (2014). Transatlantic Conflict Management Inside-Out: The Impact of Domestic Norms on Regional Security Practices. *Cambridge Review of International Affairs, 27*(2), 339–361.

Kreps, S. (2008). When Does the Mission Determine the Coalition? The Logic of Multilateral Intervention and the Case of Afghanistan. *Security Studies, 17*(3), 531–567.

Kubálková, V. (Ed.). (2001). *Foreign Policy in a Constructed World*. London: M.E. Sharpe.

Lakatos, I. (1978). *The Methodology of Scientific Research Program*. Cambridge, UK: Cambridge University Press.

Lambeth, B. S. (2001). *NATO's Air War for Kosovo: A Strategic and Operational Assessment*. Santa Monica: RAND Cooperation.

Lombardi, B. (2008). All Politics Is Local: Germany, the Bundeswehr, and Afghanistan. *International Journal, 63*(3), 587–605.

Lorenz, F. M. (1995). Forging Rules of Engagement: Lessons Learned in Operation United Shield. *Military Law Review, 75* (November/December), 17–25.

Marten, K. (2007). Statebuilding and Force: The Proper Role of Foreign Militaries. *Journal of Intervention and State-Building, 1*(2), 231–247.

Mello, P. A. (2014). *Democratic Participation in Armed Conflict*. Basingstoke: Palgrave Macmillan.

Meyer, T. (2013). Flipping the Switch: Combat, State-Building, and Junior Officers in Iraq and Afghanistan. *Security Studies, 22*(2), 222–258.

Morelli, V., & Belkin, P. (2009). *NATO in Afghanistan: A Test of the Trans-Atlantic Alliance*. Washington: Congressional Research Service. https://fas.org/sgp/crs/row/RL33627.pdf.

NATO. (2006). NATO Boosts Efforts in Afghanistan. 28 November. http://www.nato.int/docu/update/2006/11-november/e1128a.htm.

Neack, L. (2013). *The New Foreign Policy: Complex Interactions, Competing Interests*. Boulder, CO: Rowman & Littlefield.

Noetzel, T., & Rid, T. (2009). Germany's Options in Afghanistan. *Survival, 51*(5), 71–90.

Noetzel, T., & Scheipers, S. (2007). *Coalition Warfare in Afghanistan*. Briefing Paper, Chatham House. http://www.comw.org/warreport/fulltext/0710noetzel.pdf.

Noetzel, T., & Schreer, B. (2009). Does a Multi-tier NATO Matter? The Atlantic Alliance and the Process of Strategic Change. *International Affairs, 85*(2), 211–226.

Palin, R. H. (1995). *Multinational Military Forces: Problems and Prospects.* Adelphi Papers 294. Oxford: Oxford University Press.

Parker, G. (Ed.). (2005). *The Cambridge History of Warfare.* New York: Cambridge University Press.

Phillips, G. R. (1993). Rules of Engagement: A Primer. *The Army Lawyer,* July(4), 4–27.

Putnam, R. D. (1988). Diplomacy and Domestic Politics: The Logic of Two Level Games. *International Organization, 42*(4), 427–460.

Richter, A., & Webb, N. (2014). Can Smart Defense Work? A Suggested Approach to Increasing Risk- and Burden-Sharing Within NATO. *Defense and Security Analysis, 30*(4), 346–359.

Riley, J. P. (2007). *Napoleon and the World War of 1813—Lessons in Coalition War-Fighting.* London: Routledge.

Ringsmose, J. (2010). NATO Burden-Sharing Redux: Continuity and Change After the Cold War. *Contemporary Security Policy, 31*(2), 319–338.

Ringsmose, J., & Thruelsen, P. D. (2010). NATO's Counter-Insurgency Campaign in Afghanistan: Are Classical Doctrines Suitable for Alliances? *U N I S C I Discussion Papers, 22,* 56–77.

Ruffa, C., Dandeker, C., & Vennesson, P. (2013). Soldiers Drawn into Politics? The Influence of Tactics in Civil–Military Relations. *Small Wars & Insurgencies, 24*(2), 322–334.

Rynning, S. (2012). *NATO in Afghanistan—The Liberal Disconnect.* Stanford, CA: Stanford University Press.

Saideman, S. M., & Auerswald, D. P. (2012). Comparing Caveats. *International Studies Quarterly, 56*(1), 67–84.

Sky, E. (2007). Increasing ISAF's Impact on Stability in Afghanistan. *Defense and Security Analysis, 23*(1), 7–25.

Snidal, D. (1985). Coordination Versus Prisoners' Dilemma: Implications for International Cooperation and Regimes. *American Political Science Review, 79*(4), 923–942.

Soeters, J., & Manigart, P. (Eds.). (2008). *Military Cooperation in Multinational Peace Operations—Managing Cultural Diversity and Crisis Response.* London: Taylor & Francis.

Soeters, J., von Fenema, P. C., & Beeres, R. (Eds.). (2010). *Managing Military Organizations—Theory and Practice.* Oxon: Routledge.

Stoler, M. A. (2000). *Allies and Adversaries: The Joint Chiefs of Staff, The Grand Alliance, and U.S. Strategy in World War II.* Chapel Hill: University of North Carolina Press.

Trønnes, O. (2012). *Mapping and Explaining Norwegian Caveats in Afghanistan from 2001 to 2008.* Master-thesis in Political Science, Department of Sociology and Political Science, Norwegian University of Science and Technology (NTNU), Trondheim.

United Nations. (2000). *Report of the Panel on United Nations Peace Operations* (The "Brahimi Report") (A/55/305-S/2000/809). New York, NY: United Nations.

United States Department of Defense/DoD. (2005, February 9). *National Caveats' Among Key Topics at NATO Meeting.* http://www.defense.gov/news/newsarticle.aspx?id=25938.

Van der Meulen, J., & Kawano, H. (2008). Accidental Neighbours: Japanese and Dutch Troops in Iraq. In J. Soeters & P. Manigart (Eds.), *Military Cooperation in Multinational Peace Operations. Managing Cultural Diversity and Crisis Response* (pp. 166–179). Oxon: Routledge.

Von Clausewitz, C. (1976 [1832]). *On War.* Oxford: Oxford University Press.

Von Hippel, K. (2000). *Democracy by Force—US Military Intervention in the Post-Cold War World.* Cambridge: Cambridge University Press.

Weitsman, P. A. (2014). *Waging War: Alliances, Coalitions, and Institutions of Inter-State Violence.* Stanford: Stanford University Press.

Webber, M., & Smith, S. (2002). *Foreign Policy in a Transformed World.* Harlow, Essex: Prentice Hall.

Williams, M. (2011). *The Good War—NATO and the Liberal Conscience in Afghanistan.* New York: Palgrave Macmillan.

Young, T.-D. (2003). The Revolution in Military Affairs and Coalition Operations: Problem Areas and Solutions. *Defense and Security Analysis, 19*(2), 111–130.

Zinni, A. C., & Lorenz, F. M. (2000). Command, Control, and Rules of Engagement in United Nations Operations. In J. N. Moore & A. Morrison (Eds.), *Strengthening the United Nations and Enhancing War Prevention* (pp. 203–249). Durham, NC: Carolina Academic Press.

PART II

Conceptualizing Caveats

CHAPTER 2

Caveats: A Case of What?

We initially defined caveats as national reservations on the use of force in the context of coalition operations. Before elaborating on the more precise and measurable construct of caveats, the present chapter addresses the fundamental question as to *what the study of caveats might be a case of.* In asking this question, we imply, first, that our research ambition is not limited to the idiographic study of some particular caveats-case (Platt 2007: 104). The fundamental question signals that we are about to engage in the explanation of the *generic* (nomothetic) phenomenon of caveats (George and Bennett 2005: 92).

In asking, "what is caveats a case of," we also call attention to the different analytical contexts within which the politics of caveats might be explored. Hence, for the researcher, it is crucial not only to know exactly where the phenomenon of caveats starts, and where it ends but also to be aware of what broader analytical context caveats may be part of. This is a strategic choice with consequence for what research question be asked, what theoretical approaches be relevant, and what research designs may give more epistemological traction.

We have already shown our hand by indicating that caveats is a foreign policy instrument used to deal with decision-making problems related to alliance politics, domestic politics, and the politics of implementation in a political and operational environment where stakes might be extremely high. In this analytical context of generalization, we identified Foreign Policy Analysis (FPA) as a multi-level and decision-making

© The Author(s) 2019
G. Fermann, *Coping with Caveats in Coalition Warfare,*
https://doi.org/10.1007/978-3-319-92519-6_2

approach capable of encapsulating such a broad generic conception of caveats.

Below, we broaden the range of analytical gateways, and briefly discuss how caveats are understood in terms of;

- one of the several choices on a foreign policy-making decision-tree;
- one of several foreign policy instruments available for decision makers;
- one of several institutional checks on the use of force in foreign policy; and
- one of several measures available for national decision-makers to exert political control over military implementation in coalition operations.

This exercise speaks to the need to be conscious about what larger class of phenomenon we decide our subject under scrutiny to be part of. If for no other reason to extract some clues as to what generic research literature we may learn from and contribute to.

CAVEATS IN THE CONTEXT OF A FOREIGN POLICY DECISION-TREE

We may frame the phenomenon of caveats as part of a foreign policy decision-making process, resembling a decision-tree. Imagine that an alliance partner is challenged with an insistent invitation from the leader of the alliance to contribute to the establishment of a coalition force designed to counter some common threat. Assume further that the addressed alliance-partner is lukewarm towards the invitation to participate with military personnel. The hesitation may have one or several sources; a lesser sense of an immediate threat; military overstretch due to participation in other coalition operations, or military needs at home; an opposition in parliament critical to participation in out-of-area operations; a coalition government split in the middle over the issue; and so on.

On the other hand, the addressed junior partner in the alliance cannot allow itself to completely ignore the request because of its ultimate dependence on the security alliance for long-term security. The alliance is a security guarantor, and the alliance and its leader, in particular, need to be treated with respect and consideration by the other alliance partners.

This dependence is one of two main concerns that constitute the individual security dilemma in alliance politics (Snyder 1984).

On the level of individual member states, decision-makers manage the dilemma by the pragmatic balancing of competing concerns about fear of entrapment and fear of abandonment. They seek ways to optimize the dilemma of either be entrapped in a dangerous and politically risky coalition operation without any immediate national interests being served. Or to fail the solidarity clause of the alliance, and thus increase the risk of being exposed to critique in public, and even being abandoned by the alliance when the need for assistance arises in the future.

Consider the decision tree in Fig. 2.1. It shows how caveats, along with other policy decisions and options, may offer yet another means of balancing and fine-tuning a coalition partner's short-and long-term interests in alliance politics. Attaching caveats to a particular force contribution (step 4) is a foreign policy decision subtracting from the initial and more fundamental decisions to participate militarily in the coalition force (steps 2 and 3).

Alliance junior-partners do not make the initial decision to establish a coalition force (step 1). This principal decision is for great powers to make. The alliance leader initiates the establishment of the coalition force single-handedly (the US), or in consultation with the more important members of the alliance (Great Britain, France, Germany, Italy)

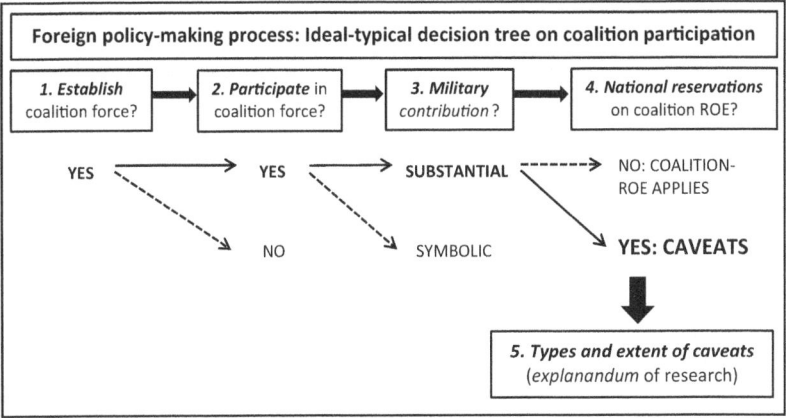

Fig. 2.1 Decision tree—contextualizing caveats

depending on circumstances. The medium and small-state members of the alliance are usually invited to participate in the decision-making process after the decision to establish the coalition force has already been made.

Typically, alliance members are invited to participate in the coalition in some military capacity (step 2). Any mode of participation will have some real political significance for the coalition. Still, there is a choice to be made for all alliance members between a symbolic or substantial military contribution (step 3). A token military contribution may take the shape of minor logistical support, the furnishing of behind-line functions, such as military hospital, or the dispatch of a handful of military staff personnel.

The offering of a substantial military contribution to the coalition effort will usually include forces capable of combat and frontline service in a considerable capacity. It may consist of the dispatch of some aircraft; some company, battalion, or division of infantry; special-forces units; tank regiments; or navy units. While such forces are supposed to fight according to some common coalition RoE, the option of national reservations on the use of force (caveats) (step 4) remains a policy instrument for the fine-tuning of the balance between concerns of entrapment and abandonment in foreign policy-making.

Such described, the study of caveats may be considered a case study of how to cope with the alliance dilemma when states' are mobilized to form coalition forces. For foreign policy-makers, this includes both the balancing of immediate and long-term concerns, as well as the weighing of national interests against the political necessities offered by the external security environment. We will elaborate on this theoretical line of reasoning in Chapter 8. The point to be made in the present chapter is that the decision on whether and how to apply caveats is separate from the decisions to participate in the coalition and what forces to bring.

CAVEATS IN THE CONTEXT OF THE BROADER RANGE OF FOREIGN POLICY INSTRUMENTS

There is a context to caveats relating to the broad repertoire of foreign policy instruments available for the state in projecting and defending national interests in global politics. These tools of foreign policy range from instruments of diplomacy, via political rhetoric and other means of political communication, and economic incentives (sticks and carrots), to clandestine operations and the use of force (Holsti 1995).

Some instruments of foreign policy-making are related to information exchange and persuasion; others include softer or harder means of enforcement.

There are ladders of escalation and de-escalation at work, both within each category and across the main dimensions of policy instruments. In describing states' relationship toward one another, it may be useful to account for the mix and calibration of foreign policy instruments involved. A lot of friendly talk and trade are obviously more benign than threats of economic boycott and the use of force (Fermann 2013: 72–83).

How do caveats relate to these foreign policy instruments? The most obvious observation is that caveats fit into the most offensive category of foreign policy tools, related to the use of lethal force. Caveats are an instance or a subcategory of the foreign policy instrument of going to war as a continuation of politics by other means. The essential moderating part is that caveats mainly is about restricting the own use of force (but as will be shown, not exclusively so) within the context of a coalition force. Recall that attaching caveats to a particular force contribution is a foreign policy decision that usually subtracts from the initial and more fundamental decision to participate militarily.

While caveats fit in mainly as a moderating element in the category of military force, caveats may relate to other instruments of foreign policy in various ways. Caveats may be more or less transparently communicated through diplomatic channels (or at the least in operational practice), and in multilateral negotiations on burden sharing. Just as we have seen instances of alliance partners paying themselves out of military commitments altogether (Japan comes to mind), side-payments of different sorts may be offered (tacitly or overtly) to compensate the alliance for the inconveniences resulting from national reservations on the use of force in coalition forces. Furthermore, a caveats-applying nation may point out military investments in other coalition operations to justify a less than full commitment in the present coalition force and so on.

The possibilities of making caveats part of the broader mix of foreign policy instruments are thus extensive. Depending on research question and approach, it may become necessary to map the calibration of this mix and the problems this bag of policy measures may have been an answer to for the foreign policy decision-makers. Indeed, as will be reasoned in

Chapters 6 and 7 the approach of FPA invites scholars to study the instrumental functions of caveats in foreign policy, as well as in domestic politics. One way of doing this is to conduct cross-level research as to whether there is a pattern of empirical correlation between a state's application of caveats, and the way the state rhetorically justifies its military contribution toward its citizens. If so, what can the reason be for the observed pattern?

The decision to use caveats, and the ability to explain the military contribution to the public, may be just what it takes to create a winning domestic political coalition for participation, and at the same time, the necessary minimum to still be considered a useful and loyal ally. In Chapters 8–10, we bring in clusters of theories from three different branches of Political Science to theorize on this multi-level reality and argue that the instrument of caveats may increase what is politically feasible to achieve in alliance politics.

This line of reasoning makes it possible to see the phenomenon of caveats in another light than the prevailing one; not merely as "a cancer that eats away at the effective usability of troops" (Jones 2004), but also as a mechanism for mobilizing more allies for the cause that would otherwise have been possible. The one thing a government of an allied nation cannot be asked to do is to adopt a controversial foreign policy that will force it to resign for domestic political reasons.

CAVEATS IN THE CONTEXT OF INSTITUTIONAL CHECKS ON THE USE OF FORCE

The institutionalist literature in security studies offers insights on domestic institutional restrictions on the use of force in foreign policy (Reiter and Stam 2002; Maoz and Russett 1993; Morgan and Campbell 1991). Institutional checks and balances on politics are yet another interesting analytical context for the study of caveats. The typology suggested by Patrick A. Mello (2015: 6) distinguishes between structural, procedural, and operational restrictions on the use of force. Caveats belong to the latter category of institutional mechanisms.

Structural restrictions include regulations on the use of force in International Law and international organizations such as the United Nations and NATO. Also, domestic institutions, not least political constitutions, belong to this category in that such political constructs may put limitations on the use of force beyond self-defense (Jakobsen 2006; Ku and Jacobson 2003; Nolte 2003).

Procedural restrictions on the use of force abroad relate to the extent to which parliament has a hand in the decision-making process, and may be able to influence decision-making both as to whether and how military force is to be applied in foreign policy (Peters and Wagner 2011).

Of considerable relevance for the political study of caveats is the final category of operational restrictions, which Mello equals to caveats (2015: 8). Mello's particular research project is to scrutinize whether there are any causal relationships between the three different kinds of limitations as to whether and how states' limit their use of force in foreign policy. For instance, whether procedural restrictions on the use of force tend to correlate with the application of operational limitations (caveats) on the use of force, and why.

For our purposes, this typology on restrictions on the use of force in foreign policy is yet another analytical context within which we may legitimately study the politics of caveats. In Mello's framework, caveats are part of the broader institutional study of political and legal restraints on the use of force in foreign policy-making. Caveats belong to the category of operational controls, which is about how force is applied, and not as structural and procedural restraints, which are mainly about whether to contribute to the war effort in the first place. In this sense, Mello's analytical framework also speaks to the decision-tree discussed above (see Fig. 2.1).

As will be shown in succeeding chapters, the approach of FPA is, with the assistance of particular explanatory theories in Political Science, capable of accounting for both international regulations, and the domestic political institutions influencing the propensity to apply caveats in foreign policy. Whether we decide to treat institutions as structures or agents, will depend on the particular research question and the institutions in question.

CAVEATS IN THE CONTEXT OF NATIONAL CONTROL OF THE USE OF FORCE

There is a final analytical context of more immediate operational relevance to the study of caveats discussed in the research literature. It relates to the exercise of national control of the contingents dispatched to coalition forces. At this point, the decision to contribute with a contingent to the coalition force has already been made. Now the time has come for national decision-makers (the principal) to decide how the powers delegated to the contingent commander in the field (the agent) are to be surveilled and controlled. For the researcher, the challenge is to figure out how to contextualize and conceptualize the instruments for national political control of the sword in a coalition context.

David P. Auerswald and Stephen M. Saideman's notion of "national controls" on the use of force (2014: 5–12) resembles Mello's category of "operational limitations" (Mello 2015). However, the two analytical constructs are not identical. For reasons related to differences in research purpose, Auerswald and Saideman (2014) need a more fine-grained construct describing the relationship between political decision-makers and operational implementers than does Mello. While Mello may do with equalizing "operational limitations" more crudely with caveats, Auerswald and Saideman distinguish between several political instruments (caveats included) which the political principal may use to control the military agent in the field.

The first main category of instruments for political control is "limits on deployed troops," and include besides "caveats," the "call home"-requirement, the on-site "red card-holder" function, and the possibility of "withholding military capabilities" from use in coalition force.

Auerswald and Saideman define caveats as "restrictions placed upon a contingent anticipating what they will be asked to do and setting rules for these circumstances" (2014: 6). Note that the term "restrictions" (rather than reservations) indicate that *permissive* caveats are not part of the concept.

The "call home"-requirement implies that the contingent commander has to ask for permission to act up the national chain of command if requested by the Force Commander to do something offensive, risky, or out of the ordinary. Even if permission is granted from the government, this may be too late for the Force Commander to utilize the contingent in question.

As to the "red card-holder" gate-keeping mechanism, this instrument of national control gives the highest national military officer within the multinational chain of command the authority to interpret the "national RoE," and decide whether and how the national contingent be used in a critical situation, or in a controversial way.

"Withholding military capabilities" from the deployment area of the coalition force is a final option for a government that needs to limit its exposure to political or military risk. Against this, we may argue that this particular subcategory does not belong in the category of "limits on deployed troops" since withheld capabilities, strictly speaking, may not yet be in the area of operation. Although politically motivated, the decision to withhold military capabilities arguably rather relate to the initial decision as to what forces to contribute (see Fig. 2.1). However, a more reasonable interpretation of Auerswald and Saideman's conception of "withholding military capabilities" from use in coalition force is to refer to the contingent which already is in the field and which cannot be used by Force Commander in certain places, for certain tasks, or in the middle of the night.

The second main category in Auerswald and Saideman's typology of national controls on the use of force in coalition forces is "oversight of deployed units" (2014: 9–10). The control effect of "caveats," "red card-holders," and "call home"-requirements is likely to be short-circuited by the failure to monitor what happens in the field and the multinational chain of command. There is rarely delegation without some oversight, as will be discussed in a subsequent chapter.

The third category relates to the poltical principal's ability to put in place "incentives for correct behavior" so military agents do not exceed their authority to make decisions on the use of force. The prospects of punishment may deter incorrect behavior, and rewards may strengthen rule-obeying behavior.

The final category in Auerswald and Saideman's typology of national control mechanisms is "selecting military commanders." Selecting the right staff and military leaders are half the job done from the perspective of the political principal. Where political sensitivities are internalized in the officer responsible for making difficult decision in the field, strong incentives related to reward and punishment are likely to be less necessary to rely upon (Auerswald and Saideman 2014: 10–12).

For the study of the politics of caveats, Auerswald and Saideman's typology treats caveats in the context of how nations may insert some political control over the national contingent in coalition forces. Research on national control of military implementation is a key part of the civil–military relations field of research, which applies principal-agent theories to make sense of relationships between political decision-makers and military agents (Huntington 1957).

With no claim for being exhaustive, the purpose of the chapter has been to demonstrate how we may analytically contextualize caveats in different ways and for different research purposes. In the process, we have also provided clues to subsequent chapters in that contextualization of caveats is probably the best preparation for the subsequent conceptual delimitation of the concept. To make an informed decision on the concept and subclasses of caveats, it is useful to know what caveats might be a subclass of itself.

REFERENCES

Auerswald, D. P., & Saideman, S. M. (2014). *NATO in Afghanistan: Fighting Together, Fighting Alone*. Princeton, NJ: Princeton University Press.

Fermann, G. (Ed.). (2013). *Utenrikspolitikk og norsk krisehåndtering*. Oslo: Cappelen Damm Akademika.

George, A., & Bennett, A. (2005). *Case Studies and Theory Development in Social Sciences*. Cambridge, MA: MIT Press.

Holsti, K. J. (1995). *International Politics: Framework for Analysis*. Englewood Cliffs: Prentice Hall.

Huntington, S. P. (1957). *The Soldier and the State: The Theory and Politics of Civil–Military Relations*. Cambridge, MA: Belknap Press.

Jakobsen, P. V. (2006). *Nordic Approaches to Peace Operations: A New Model in the Making?* Oxon: Routledge.

Jones, J. L. (2004). Prague to Istanbul: Ambition Versus Reality. Global Security: A Broader Concept for the 21st Century. Center for Strategic Decision Research 21st International Workshop on Global Security—Berlin, 7–10 May. http://csdr.org/2004book/Gen_Jones.htm.

Ku, C., & Jacobson, H. K. (2003). Broaching the Issues. In C. Ku & H. K. Jacobson (Eds.), *Democratic Accountability and the Use of Force in International Law* (pp. 3–35). Cambridge: Cambridge University Press.

Maoz, Z., & Russett, B. (1993). Normative and Structural Causes of Democratic Peace, 1946–1986. *The American Political Science Review, 87*(3), 624–638.

Mello, P. A. (2015). Constitutional and Political Restrictions on the Use of Force: A Comparative Analysis of Prevalent Regulations and Their Evolution Over Time. Paper, International Studies Association's 55th Annual Convention, Toronto, 26–29 March.

Morgan, T. C., & Campbell, S. H. (1991). Domestic Structure, Decisional Constraints, and War: So Why Kant Democracies Fight? *Journal of Conflict Resolution, 35*(2), 187–211.

Nolte, G. (Ed.). (2003). *European Military Law Systems*. Berlin: Gruyter.

Peters, D., & Wagner, W. (2011). Between Military Efficiency and Democratic Legitimacy: Mapping Parliamentary War Powers in Contemporary Democracies, 1989–2004. *Parliamentary Affairs, 64*(1), 175–192.

Platt, J. (2007). Case Study. In W. Outhwaite & S. P. Turner (Eds.), *The Sage Handbook of Social Science Methodology* (pp. 102–127). London: Sage.

Reiter, D., & Stam, A. C. (2002). *Democracies at War*. Princeton, NJ: Princeton University Press.

Snyder, G. G. (1984). The Security Dilemma in Alliance Politics. *World Politics, 36*(4), 461–495.

CHAPTER 3

Regulation of the Use of Force in Military Organization and Coalition Forces

Why do governments reserve how the military can fight in coalition forces, when it may put soldiers in harm's way, and make it more difficult to accomplish the mission? Why issue caveats when they are potentially damaging to the military efficiency and political effectiveness of the coalition?

The first step in any attempt to provide qualified answers to such topical research questions, is to construct an analytical language capable of precisely describing the phenomenon under scrutiny. We initially defined caveats as national reservations on the use of force in the context of coalition operations on the assumption that the concept needs further elaboration. The definition was half-baked because it lacked both analytical and material context, as well as specification of conceptual properties and operational definitions necessary to engage in empirical research. In the previous chapter, we sketched out four different and partly overlapping analytical contexts in which to grasp caveats as national reservations on the use of force. This we did on the argument that it is useful to know what caveats be a sub-class of—for theoretical as well as conceptual reasons.

In the present chapter, we prepare for the informed decision on how to define and operationalize the concept of caveats by first consulting the empirical record on coalition collaboration relating to deviations in the use of force briefly touched upon in the introductory chapter. Next, we tap into the literature on rules of engagement (RoE) as a mechanism for the regulation of the use of force in a military organization—multinational

© The Author(s) 2019
G. Fermann, *Coping with Caveats in Coalition Warfare*,
https://doi.org/10.1007/978-3-319-92519-6_3

military operations included. Subsequently, RoE be used as a key yardstick against which caveats be measured. For the time being, we reserve the term "rules of engagement" (RoE) for the body of principles and rules that supervise soldiers and commanders in the use of force. RoE vary in robustness, specificity, and the extent to which military personnel granted the discretion to interpret guidelines for the use of force.

Reviewing the recent history of caveats-use and discussing the finer points of regulation of force on the ground represent the bottom-up approach to concept development. In Chapter 11, we discuss this move within the case design of the inductive plausibility probe. In the generalizing plausibility probe, we juxtapose "facts on the ground" with promising but tentative analytical constructs, such as some preliminary conception of caveats. This procedure allows us to "refine the operationalization or measurement of key variables" to prepare for generic research (Levy 2008: 6). In learning from the operational and military experience, we hope to be in a position to reason a conception of caveats that are recognizable to military professionals and discrete and nuanced enough to be useful in empirical research.

A Brief History of Caveats as National Deviations from the Coalition Rules of Engagement

As noted in the introductory chapter, the synonymous terms of "national RoE" and "caveats" came into use with the need to describe the growing challenges with coordinating the use of force in UN peace operations in the 1990s, and in NATO's operation in Afghanistan from 2003. With the end of the Cold War, there was a tremendous increase both in the number of civil wars and ethnic violence, but also in states' willingness to send armed forces to intervene in these conflicts (Rost and Greig 2011). The scope of the operations was so broad as to include the providing of security, disarmament, demobilization, reintegration, support for humanitarian and human rights assistance, and governance.

This demanding portfolio of functions implied that most national armed forces were not capable of conducting missions alone. To cope with the situation, states had to pool military resources and work together in ad hoc multinational military coalitions led by the UN, NATO, EU, or major powers like Great Britain, France, and the United States in particular. During the 1990s, UN operations in, for example, Somalia, Bosnia, Haiti, Rwanda, East-Timor, Kosovo, Congo, and Sierra Leone taught the international community that the requirements of these missions went far beyond what the coalition forces were prepared for and capable of doing. One of the problems that surfaced with these operations was the escalating use of force in the areas of operation (Findlay 2002: 124–314).

To avoid taking the side and become an active part in the conflicts, the intervening UN military forces often operated with restrictive (un-robust, weak) coalition RoE, sometimes to the extent that they failed to sufficiently act in self-defense, or complete their missions. The gravest and most publicized examples were the failure of on-site UN-soldiers to use force to prevent the 1994 genocide in Rwanda, and the 1995 massacres in Srebrenica in Bosnia-Herzegovina. As a result, military commentators and analysts started to analyze the political and operational limitations of traditional UN operations concerning the use of force and RoE (Berkowitz 1994; Fair 1997; Klep and Winslow 1999; Lorenz 1995; Reed 2000; Zinni and Lorenz 2000).

While it was clear in the aftermath of these operations that the coalition RoE often was too restrictive for the violent circumstances developing, analysts also observed national differences within the coalitions in how states' forces applied the coalition RoE. As noted in the introductory chapter, different interpretations of the coalition RoE were at the time labeled "national RoE" and referred to how different nations understood the same rules for the use of force in particular ways (Bennett and MacDonald 1995; Miller 1995: 19; Palin 1995: 7; Phillips 1993: 24).

Because mandates issued by the UN Security Council often were formulated in a general, diplomatic language, and because the UN lacked expertise and resources in military planning, concepts of operations and RoE were often unspecified, and consequently ambiguous. This state of affairs was an invitation to contributing nations with divergent political agendas and operational constraints to read different things into how the coalition RoE be practiced in the field. Without any explicit clarification, each nation often interpreted and operationalized relevant terms like "self-defense," "imminent threat," and "hostile act" quite differently (Findlay 2002: 351–359). As noted, national differences in military training and culture also played a role in bringing about divergent interpretations of coalition RoE, as was national differences in the legal status of RoE.

Notwithstanding the challenges related to different interpretations of common coalition RoE, it did not cause significant controversy compared to other problems concerning the use of force—such as deficiencies in equipment, lack of relevant training and experience of troops, poor command and control arrangements, and lack of intelligence-gathering capabilities. Also, the geographical segregation of forces into national sectors of deployment implied that there was the scant need to cope with national differences in the understanding and practicing of coalition RoE (Palin 1995: 18). Direct work-related contact between personnel from various national contingents occurred only at the level of headquarters, thus causing only minor difficulties for coordinating operations (Lorenz 1995: 21–22).

The issue of diverging "national RoE" became significantly more critical in the typical ad hoc "coalition-of-the-willing" operations after the September 11, 2001 attacks on the United States. In these operations, different national units with specialized capabilities were working side by side in the field and frequently integrated into bi- or multinational contingents. This pattern of military integration on the ground was certainly the case with NATO's ISAF-operations in Afghanistan (2003–2014) and was not easily reconcilable with the fact that member-states in ISAF held different views on the nature of the operations. The inability to reach consensus on military strategy was rooted in what some governments wanted to make into a counter-insurgence campaign, while other governments preferred to see as a stabilization and reconstruction mission. This fundamental disagreement came to the surface in national differences not only with the number of troops and what kind of military capabilities states' participated with but also on *how* states took part in the operations regarding the use of force. With different national units specialized in different capabilities working together on the tactical level, differences in "national RoE" were quickly exposed in the field in ways that were difficult to cope with.

The problem with national differences in the commitment to use force was, as mentioned, pushed to the fore in 2005 when the low-intensity conflict in Afghanistan exploded into a full-fledged insurgency (Cimbala and Forster 2010: 150–153). Differences in the willingness to use force in response to this development were frequently referred to as "national restrictions," or "caveats" (Kreps 2008: 562; Marten 2007: 243; de Nevers 2007: 51; Ringsmose 2010: 328). These were reservations to the common coalition RoE, and showed in many forms: Some states' military units were restricted from engaging in offensive operations. Others needed political clearance from national political authorities before accepting orders from NATO. Some nations refused to allow NATO deploy their forces to areas with high levels of violence and risk for own casualties (Auerswald and Saideman 2012, 2014; Saideman and Auerswald 2012).

Because these caveats challenged states' ability to operate side by side in the field seamlessly, it was not just a matter of sorting out national differences in headquarters as in previous UN operations. As the different national reservations surfaced, they complicated NATO's military planning and operational efficiency. For such reasons, caveats became controversial among coalition members who were concerned about military efficiency and burden sharing early on in the ISAF-operation (Noetzel and Schreer 2009). Recall that the problem with caveats was high on the agenda at NATO's Riga-conference in 2006. During the conference all members promised to review their use of caveats, but with mixed results (Cimbala and Forster 2010: 154). Caveats continued to challenge both NATO's operational flexibility and the cohesion among coalition partners throughout the Afghanistan engagement.

This may suffice to demonstrate how "caveats" and "national RoE" grew out of similar challenges with harmonizing and coordinating diverging national RoE in coalition operations. The two concepts of national RoE and caveats are different terms but refer to similar if not identical phenomenon: *national reservations on the use of force as deviating from the common coalition RoE*. Before specifying the particular properties of this crude definition of caveats, we need to address the architecture and functions of RoE in any military organization, multinational or national. If not being completely undermined by caveats, common RoE is established to overcome coordination problems in military operations, secure that the political mandate of the mission is properly translated to military conduct in the field, and to ensure the effective implementation of military operations.

RULES OF ENGAGEMENT IN MILITARY OPERATIONS

When states join forces in a multilateral military operation, they pool military units and capabilities and organize their forces under a single unified command. For the different national military units to work effectively under a unified command, all forces need to follow the same rules and modus operandi. Otherwise, they will have to operate parallel to each other, each under its national command, and it will be difficult—if not impossible—to coordinate action among them. Still, in military coalitions, states frequently issue caveats on how the coalition command can use their units.

Such caveats are national reservations to the common coalition ROE, and typically restrict a national military unit's ability to conduct parts of an operation. National reservations may take different forms, including geographical restrictions on the use of a unit; requirement for additional information on the nature and identity of targets before striking; or higher risk-reluctance as to what loss of own life or civilians casualties accepted when operating. Coalition RoE is thus a key benchmark against which national deviations in the use of force in coalition operations be measured. Prior to specifying the full definitional properties and operational dimensions of the concept of caveats, we need to go into some detail on the operational and political functions of RoE in military organization, and the conflicting concerns the mechanism of RoE and the delegation of authority to make decisions on the use of force are intended to balance.

In delivering on this task, we make liberal use of Per Marius Frost-Nielsen's study "Towards a Holistic Understanding of Rules of Engagement," which is to be published in *Journal of Military Ethics* (2018). In a field dominated by a legal literature that conceive of RoE as an instrument keeping the execution of war within the ramification of International Law (Jus in Bello), Frost-Nielsen's contribution relates to the *political* function of RoE keeping war the continuation of politics by other means, and the *operational* need to coordinate use of force in military organization. In the effort to prepare the concept of caveats for empirical research, such angles to RoE are highly relevant.

Rules of Engagement as Guidelines for the Use of Force

Elaborating on Alexander L. George (1991) and Scott Sagan (1991), Frost-Nielsen defines RoE as "guidelines stating (i) the conditions needed to be met before (ii) specified actions—including the use of lethal force—can be carried out" (2018: 8). The RoE's condition-action rules for the use of force addresses "*when, where,* against *whom,* and *how* military force be used," and "*who* has the authority to make a judgment about the conditions and approve actions" on use of force (ibid.). Whose decision-making challenges is RoE addressing in the field, and what particular operational problems may RoE give guidance to manage?

Frost-Nielsen notes that soldiers come in different shapes, and is located at all levels in the chain of command. It may be the soldier guarding a control post or a fighter pilot in a cockpit patrolling the airspace. The RoE may supervise a "navy captain on the bridge with the command of all of the vessel's weapon systems," or any other personnel with "authority to make a judgment about the use of military force." What soldiers across branches and levels share is the need for some guidance that "informs them about the conditions that need to be present before using force" (2018: 8).

For instance, are the soldier allowed to "fire at anyone approaching his control post, or only when those approaching do not halt to warning shots"? Does the soldier have to ask his superiors for permission to fire, or does he have "the authority to make these decisions himself" (ibid.)?

Does the fighter pilot have the authority to "engage another aircraft simply from the information of his radarscope, or is he required to visually confirm that it is an enemy aircraft before firing missiles at it"? How does he visually distinguish friend, foe and neutral aircraft on the horizon? "From the aircraft's insignia or its aggressive maneuvers"? Indeed, what constitutes "aggressive maneuvers" (ibid.: 9)?

Is the navy captain allowed to fire the first shot, or does he have to wait "for an enemy vessel to fire the first shot before he can decide to attack the enemy in self-defense"? At what range is he allowed to fire at "an enemy vessel in self-defense"? Does it make a difference if the situation "takes place in territorial or international waters" (ibid.)?

RoE is guidelines supposed to make soldiers know when to fire their weapons, and when not to. Ideally, the RoE should offer precise guidance to handle the situations described above. However, RoE is rarely a precise script capable of micro-managing the soldier's use of force in every situation. Uncertainty prevails. The decision to fire the first shot or react with overwhelming force is not always executed with full information of the enemy's intentions. Frost-Nielsen describes this decision-making situation under uncertainty as "a micro-level security dilemma of untenable choices," and argues that "RoE may never fully solve this dilemma in advance of combat" (2018: 9). However, what RoE does provide is some "support for soldiers and commanders' judgment by stating the conditions required to be present before they can fire their weapons" (ibid.).

Rules of Engagement as a Mechanism Securing that War Are the Continuation of Politics by Other Means

Setting requirements for the use of force to support operational decision-making are not the only function of RoE. By specifying the requirements for the use of force, the RoE is also designed to serve the political intentions as reflected in the mandate of the military operation.

For instance, requiring the soldier to shoot warning shots might "reduce the risk of killing innocent civilians." Such precaution is "vital if the overall political goal of the military operation is to win the hearts and minds of the local population" (Frost-Nielsen 2018: 10). To avoid the unwanted escalation of the conflict, the RoE may be written such that the fighter pilot is required to "visually identify a suspicious aircraft before engaging" (ibid.).

Furthermore, by distinguishing between territorial and international waters, the navy captain has to consider specific requirements established in international law when making decisions on the use of force. Frost-Nielsen notes that such awareness is "crucial if the reason for sailing a military vessel in the same waters as an enemy fleet is to uphold legal regulations of international waters" (ibid.).

While RoE serves political, operational as well as legal purposes, in the political study of caveats we are inclined to focus on RoE as an instrument to harness the use of military force to some political rationale. In RoE decision-makers have a tool to ensure that war continues to be a reflection of politics with other means (von Clausewitz 1976 [1832]).

Rules of Engagement as "Standard Operating Procedure"

In the previous, we have understood RoE as guidelines designed to support the operational judgment in use of force for military and political purposes. By instead conceiving of RoE as *standard operating procedures* (SOP), decision-making on the use of force reduced to a choice of picking the correct pre-programmed script for action. An SOP is a standardized "off-the-shelf" recipes for conduct modeled according to an organization's purpose, expertise and previous experience in problem-solving. The recipes prescribe how the organization should react confronted with a recurrent problem. Acting according to such recipes allows the organization to view the problem as an instance of something already known. The main argument in favor of SOP is that such decision-making recipes reduce the complexity of problem-solving in organizations in that the SOP defines how the problem should be addressed. Such *programming* of the decision-making procedure limits the scope for relevant and necessary information that need considering, as well as the number of possible solutions to choose from when deciding how to deal with the standardized problem in question.

When the overall problem is recognized and understood, it can be divided into smaller and more tangible issues, and addressed by different groups or individuals with appropriate kinds of expertise. When the organization works according to the same pre-defined sets of recipes, it is in principle possible to efficiently coordinate and perform concerted action among numerous members of the organization. In sum, SOP establishes the boundary within which bureaucrats and specialists understand the problem at hand. SOP also clarifies the limits for how decision-makers can use their expertise and judgment to find practical solutions under prevailing circumstances (De Mesquita 2006: 170–172).

For the present study, the organization is the coalition force under a unified command, and the members of the organization are commanders and soldiers. In combat, military units sustain only as long individual members commit themselves to collective goals at the cost of personal injury or even death. Maintaining necessary organizational cohesion to ensure competent military performance under conditions of danger and the imminent prospects of death is often done by executing actions learned through inculcated collective drills (King 2006: 493–495). Drills are the learning of SOPs at the most practical level of warfare, recipes used to reduce the chaotic complexity in combat and to coordinate actions among personnel.

Situational Judgment: Balancing Political and Operational Concerns

We may perceive of ROE as a military sub-class of the SOP. In military operations, RoE provides soldiers and commanders with guidelines for the use of force. Being a flexible type of SOP, RoE should not be mistaken for orders telling soldiers and commanders exactly how to use force in all and every conceivable situation. Rather, RoE is guidelines to and limits for how to *think* about a problem in a given situation, and thus a point of departure for making sound decisions as to how force may be applied to solve military challenges under particular circumstances. Indeed, combat includes "too many unpredictable and changing circumstantial factors [...] for RoE to offer precise guidance" (Frost-Nielsen 2018: 10). If RoE were to take the shape of recipes for actions, instructions or specific orders, this key mechanism for the regulation of the use of force would not be able to contribute well to the translation of political, legal, and operational requirements. Instead, the unpredictable and often confusing dynamics of combat requires an understanding of RoE as an *interpretive framework in need of competent interpretation*. This conception of RoE is captured in Mark J. Osiel's formulation that regulation of the use of force should primarily "encourage [soldiers and commanders] to exercise situational judgment on the basis of local knowledge, assessed in light of prior knowledge" (2002: 359).

Situational judgment is a key to the sound interpretation of RoE. The component of local knowledge includes military experience as relates to the management of organized violence and the execution of military action. Local knowledge is the instrumental skill of deciding whether the situation in the field warrants the use of force and how. However, since the state-organized use of force serves some political purpose (war as "politics by other means"), the situational judgment also requires the ability to take into account the wider operational and political context of the situation, using prior (or general) knowledge (Frost-Nielsen 2018: 10–11).

What do we require of RoE to assist (as opposed to instruct) situational judgment in circumstances characterized by uncertainty and the unforeseen? Frost-Nielsen argues that "RoE should ideally be general enough to give soldiers guidance relevant in all possible situations, and specific enough to provide useful support for decision-making in any given situation" (2018: 11). This assessment resonates with Scott D. Sagan's evaluation that RoE needs to be "deliberately written in a flexible manner to balance the legitimate need for top-level guidance on the appropriate action with the necessity for field-level judgments about specific conditions threats, and opportunities" (1991: 444). This more than indicate that RoE cannot be reduced to the mechanical application of the instructional text to the questions of whether and how to use force. To optimize goal-achievement related to military efficiency, legal accountability, and political effectiveness, on-scene commanders and soldiers need to apply their situational judgment as to how the letter and spirit of the RoE are to be understood in particular combat situations.

Rules of Engagement as a Guide to Operational Decision-Making: How Specific? How Much Delegation? How Robust?

Recall that RoE's condition-action rules for the use of force addresses when, where, against whom, and how (much) military force should be used, and who has the authority to make the judgment about the conditions and approve actions on use of force. Of relevance for the subsequent operationalization of caveats as national reservations on the use of force is the fact that RoE vary in three crucial regards, relating to (i) the *specificity* of RoE; (ii) the extent to which interpretation of RoE *delegated* to the military and down the chain of command; and (iii) the *robustness* of RoE in allowing for the more extensive use of force.

How detailed the RoE description of conditions and actions for the use of force is, may range on a continuum from the very general to the highly specific. We should also be open to the possibility that the conditions for *initializing* the use of force are more specified than the RoE's specification as to *how* the use of force is to be executed. Or the other way around, as this is an empirical question.

The more specific a set of ROE describes the conditions that must be met before specified actions relating to the use of military force can be carried out, the more instructive it is. The advantage of the detailed specification of conditions is that it significantly reduces the amount of information and the number of possible solutions a commander or a soldier under pressure needs to consider to make sense of the situation and make decisions about the use of force. However, when unforeseen circumstances arise, the soldier will likely benefit from the flexibility allowed for in a RoE that is formulated in a general language as to under what conditions and how force may be used. Low specificity comes with the advantage of greater autonomy for the on-scene commander to utilize his judgment to seize initiative and opportunities that might emerge in unexpected ways. The flip side to this freedom of action is the need to consider more information and possible solutions, in particular situations (ibid.: 12–13).

The second dimension of RoE relates to how delegating the RoE is concerning *who has the authority* to make judgments about conditions for using force and approve of specific military actions. As will be elaborated on in Chapter 10, authority may be delegated in two ways. The first, positive command, refers to an action that can be taken only on the direct orders from higher levels of command. The second, negative command, or command by negation, refers to actions that are taken under specified circumstances at the discretion of an officer at a certain level of military command unless negated by counter-manding orders from higher military or political authorities (George 1991: 18).

Through arrangements of positive and negative command given in ROE, decision-making authority is placed somewhere in the chain of command. The chain of command ranges from the top-level political leaders and policymakers (presidents, cabinets), through bureaucratic organizations (department of defense, joint chiefs-of-staff), to military headquarters and command posts in the field, all the way down to soldiers on the ground, in the air, or at sea. We assume the in-depth contextual understanding of a particular situation (local knowledge) be greater at the lower levels in the chain of command, while the understanding of overall political, legal, and strategic intentions (prior, general knowledge) will be increasing as we move up the chain of command (Frost-Nielsen 2018: 12).

The degree of delegation of authority given in any RoE ranges on a continuum from highly assertive to highly delegating (Feaver 1992: 7–12). The more delegating the RoE, the further down the chain of command the authority to judge conditions and approve actions is located. The more assertive the RoE, the higher up in the chain of command the authority will rest. Designing RoE more or less delegated or assertive is to decide whether local or more general prior knowledge be given priority in decision-making on the use of force (Frost-Nielsen 2018: 12). In a principal-agency perspective to be elaborated on in Chapter 10, a high degree of assertiveness reflects that the political principals want to exert closer political control with the implementation executed by military agents in the field (Auerswald and Saidelman 2014: 13–15).

The two dimensions of specificity of RoE and delegation of authority to decide on the interpretation of RoE in military operations are for most practical purposes inter-related. For instance, during NATO's operation in the 2011 Libyan conflict, fighter aircraft were bombing targets in densely populated areas. The targets were differentiated based on the estimated levels of civilian or collateral damage that might result from striking. The higher the expected collateral damage, the higher up the chain of command the authority to decide to engage was placed (De Cock 2012: 33).

Below is a typology of RoE combining the degree of specificity of RoE concerning conditions for the use of force and the particular actions allowed to execute, and the extent to which decision-making authority is delegated down the chain of command (Fig. 3.1). The North Eastern category relates to RoE with a high degree of specification of conditions for the use of force and the calibration of military actions, combined with a high degree of delegation of authority to make judgment far down the chain of command. An illustration of this category is the US air offensive against Iraq in the Gulf War in 1991. The RoE was very specific on the conditions for the use of force to avoid collateral damage. At the same time, the US military (US CENTCOM, and the command centers in the Gulf region) enjoyed considerable discretion to make operational decisions within the RoE without any interference from political decision-makers (Humphries 1992).

The South Eastern category pertains to RoE that offers a high degree of specification of conditions for and actions related to the use of force,

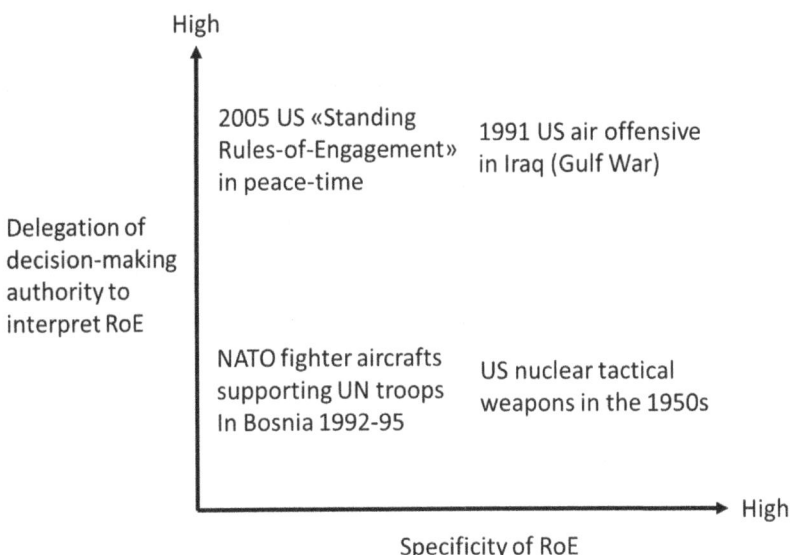

High

2005 US «Standing Rules-of-Engagement» in peace-time

1991 US air offensive in Iraq (Gulf War)

Delegation of decision-making authority to interpret RoE

NATO fighter aircrafts supporting UN troops In Bosnia 1992-95

US nuclear tactical weapons in the 1950s

High

Specificity of RoE

Fig. 3.1 Typology—rules of engagement (RoE) as a function of the degree of authority to interpret RoE and the degree of specificity of RoE concerning when, where and how to use force

combined with authority located higher up in the chain of command. Fitting this category is the US political authorities' control with the use of tactical nuclear weapons in the 1950s. The US RoE for the use of nuclear weapons was highly specified and detailed. Nonetheless, the delegation of authority further down the chain of command was applicable only under highly specified and rare circumstances (different levels of alert readiness, or Defense Condition—DEFCON). This arrangement kept the political authorities, potentially all the way up to the President, involved in the decision-making loop, for rather obvious reasons (Frost-Nielsen 2018: 15; Roman 1998: 145–160).

The South Western category accounts for RoE where conditions and actions are described in general terms, combined with a low degree of delegation of decision-making authority. Here RoE is constantly open to interpretation and involves political decision-makers and the upper echelons in the military chain of command. Arguably, this category describes the RoE of many UN operations during the 1990s. The RoE for the use of NATO's fighter aircraft in support of UN troops in Bosnia in 1992–1995 is instructive. The RoE lacked specification that made them ambiguous and subject to interpretation. Simultaneously, there was the limited delegation of authority as the use of force had to be approved at high levels in both the UN (all the way up to the Secretary-General), and in the NATO chain of command on a case-to-case basis (Reed 2000).

The final North Western category covers RoE where descriptions of conditions and actions for the use of force are formulated in a general and ambiguous language, combined with a high degree of delegation of authority (even down to commanders and soldiers at the platoon level) to decide as to how the general RoE be applied. A case in mind is the US Navy initiated system of standardized RoE for peacetime operations. Because this kind of RoE were standing, it was not designed for a particular conflict or specific types of an incident but had to be generic enough to cover any circumstance. The RoE was also delegating because the purpose was to facilitate planning and support judgment of on-scene commanders, and even individual soldiers using force for self-defense (Parks 1989: 84–86).

The typology presented above provides a foundation for observing and comparing RoE along the two key decision-making dimensions in the regulation of the use of force in military organizations. The typology allows us to distinguish between different sub-classes of coalition RoE as they may appear in particular coalition operations. Crucial for our caveats-delimiting purposes, a more thorough understanding of RoE holds the key to capturing a defining attribute of the concept of national reservations on the use of force in coalition operations since caveats on the use of force have no precise meaning if not related to a common framework regulating the use of force.

A final dimension of RoE remains to explain. RoE varies in terms of robustness across cases. Robustness describes the extent to which RoE (any RoE, coalition or national) allows for the use of force. A robust RoE reflects in the military capabilities of the force but qualifies as a separate phenomenon. The scale of robustness may stretch from the UN peacekeeping operation only allowed to use force in strict individual self-defense, to the very robust mandates of intervening enforcement operations called out to fight and win a war for some political purpose. We may crudely distinguish between robust, moderate, and weak coalition RoE, as we will reserve the use of the terms "restrictive" and "permissive" for national reservations on the use of force compared to coalition RoE, and not to describe the RoE itself.

The variation in coalition RoE as to robustness implies that any comparison of caveats across cases need to account for the probability that the caveats compared do not relate to the same coalition RoE robustness (Husby 2015: 23). This discrepancy does not represent a methodological problem when comparing different states' caveats within the same operation, but does if comparing caveats across different coalition forces, or comparing caveats in a coalition force where the robustness of coalition RoE changes with time. In the latter case, comparing absolute caveats makes less sense than comparing relative caveats adjusted for differences in coalition RoE robustness.

On this backdrop, we are finally prepared to specify the concept of caveats, so far only vaguely defined as national reservations on the use of force in coalition operations.

References

Auerswald, D. P., & Saideman, S. M. (2012). Coalitions at the Limits—NATO's Restricted Effort in Afghanistan. In P. C. McMahon & J. W. Western (Eds.), *The International Community and State-Building. Getting Its Act Together?* Oxon: Routledge.

Auerswald, D. P., & Saideman, S. M. (2014). *NATO in Afghanistan: Fighting Together, Fighting Alone.* Princeton, NJ: Princeton University Press.

Bennett, D. A., & MacDonald, A. F. (1995). Coalition Rules of Engagement. *Joint Force Quarterly*, Summer(8), 124–125.

Berkowitz, B. D. (1994). Rules of Engagement for U.N. Peacekeeping Forces in Bosnia. *Orbis*, *38*(4), 635–636.

Cimbala, S. J., & Forster, P. K. (2010). *Multinational Military Intervention: NATO Policy and Burden-Sharing*. Farnham: Ashgate.

De Cock, C. (2012). Operation Unified Protector: Targeting Densely Populated Areas in Libya. *Military and Strategic Affairs*, *4*(2), 25–35.

De Mesquita, B. (2006). *Principles of International Politics: People's Powers, Preferences, and Perceptions*. Washington: Congressional Quarterly.

de Nevers, R. (2007). NATO's International Role in the Terrorist Era. *International Security*, *31*(4), 34–66.

Fair, K. V. (1997). The Rules of Engagement in Somalia—A Judge Advocate's Primer. *Small Wars & Insurgencies*, *8*(1), 107–126.

Feaver, Peter D. (1992). *Guarding the Guardians: Civilian Control of Nuclear Weapons in the United States*. Ithaca: Cornell University Press.

Findlay, Trevor. (2002). *The Use of Force in UN Peace Operations*. Oxford: Oxford University Press.

Frost-Nielsen, P. M. (2018, forthcoming): Bringing Military Conduct Out of the Shadow of Law: Towards a Holistic Understanding of Rules of Engagement (ROE). *Journal of Military Ethics*, *17*(1–2).

George, A. L. (1991). The Tension Between "Military Logic" and Requirements of Diplomacy in Crisis Management. In A. L. George (Ed.), *Avoiding War—Problems of Crisis Management* (pp. 124–143). Boulder, CO: Westview Press.

Humphries, J. G. (1992). Operations Las and the Rules of Engagement in Operations Desert Shield and Desert Storm. *Airpower Journal*, *11*(3), 25–41.

Husby, G. (2015). *Fra hull i luften, til hull i Gaddafis bunker. Bruk av politiske reservasjoner på norsk militærmakt i flernasjonale koalisjonsoperasjoner. En komparativ studie av F-16 bidragene i Kosovo, Afghanistan og Libya, belyst gjennom utenrikspolitisk analyse*. Master Thesis in Political Science, Trondheim: Department of Sociology and Political Science, Norwegian University of Science and Technology (NTNU).

King, A. (2006). The Word of Command—Communication and Cohesion in the Military. *Armed Forces and Society*, *32*(4), 493–512.

Klep, C., & Winslow, D. (1999). Learning the Lessons the Hard Way—Somalia and Srebrenica Compared. *Small Wars & Insurgencies*, *10*(2), 93–137.

Kreps, S. (2008). When Does the Mission Determine the Coalition? The Logic of Multilateral Intervention and the Case of Afghanistan. *Security Studies*, *17*(3), 531–567.

Levy, J. S. (2008). Case Studies: Types, Designs, and Logics of Inference. *Conflict Management and Peace Science*, *25*(1), 1–18.

Lorenz, F. M. (1995). Forging Rules of Engagement: Lessons Learned in Operation United Shield. *Military Review*, *75*(6), 17–25.

Marten, K. (2007). State-Building and Force: The Proper Role of Foreign Militaries. *Journal of Intervention and Statebuilding, 1*(2), 231–247.

Miller, E. S. (1995). *Inter-operability of Rules of Engagement in Multinational Maritime Operations.* Arlington, VA: Center for Naval Analysis.

Noetzel, T., & Schreer, B. (2009). Does a Multi-tier NATO Matter? The Atlantic Alliance and the Process of Strategic Change. *International Affairs, 85*(2), 211–226.

Osiel, M. J. (2002). *Obeying Orders—Atrocity, Military Discipline and the Law of War.* Piscataway, NJ: Transaction Publishers.

Palin, R. H. (1995). *Multinational Military Forces: Problems and Prospects* (Adelphi Papers No. 294). Oxford: Oxford University Press.

Parks, W. H. (1989). Righting the Rules of Engagement. *U.S. Navy Institute Proceedings, 115*(5), 83–93.

Phillips, G. R. (1993). Rules of Engagement: A Primer. *The Army Lawyer,* July(4), 4–27.

Reed, R. M. (2000). *Chariots of Fire: Rules of Engagement in Operation Deliberate Force.* Maxwell AFB, AL: Air University Press.

Ringsmose, J. (2010). NATO Burden-Sharing Redux: Continuity and Change After the Cold War. *Contemporary Security Policy, 31*(2), 319–338.

Roman, P. J. (1998). Ike's Hair-Trigger: U.S. Nuclear Predelegation, 1953–60. *Security Studies, 7*(4), 121–164.

Rost, N., & Greig, M. G. (2011). Taking Matters Into Their Own Hands: An Analysis of the Determinants of State-Conducted Peacekeeping in Civil Wars. *Journal of Peace Research, 48*(2), 171–184.

Sagan, S. D. (1991). Rules of Engagement. In A. L. George (Ed.), *Avoiding War—Problems of Crisis Management* (pp. 443–470). Boulder, CO: Westview Press.

Saideman, S. M., & Auerswald, D. P. (2012). Comparing Caveats: Understanding the Sources of National Restrictions upon NATO's Mission in Afghanistan. *International Studies Quarterly, 56*(1), 67–84.

von Clausewitz, C. (1976 [1832]). *On War.* Oxford: Oxford University Press.

Zinni, A. C., & Lorenz, F. M. (2000). Command, Control, and Rules of Engagement in United Nations Operations. In J. N. Moore & A. Morrison (Eds.), *Strengthening the United Nations and Enhancing War Prevention* (pp. 203–249). Durham, NC: Carolina Academic Press.

CHAPTER 4

Preparing the Concept of Caveats for Empirical Research

In the Western multilateral military community, the term caveats are used to describe national restrictions on the use of force in coalition operations. In the political study of national reservations on the use of force, caveats have been used with reference to domestic decision-making over military support to coalition operations and national control of military operations (Auerswald and Saideman 2014; Saideman and Auerswald 2012), state-building missions and multinational operations (Marten 2007), multilateral military intervention and burden sharing (Kreps 2008; Ringsmose 2010), political cohesion in military alliances (Høiback 2009), and in relation to operational effectiveness in NATO operations (de Nevers 2007). The increasing use of the term in conjunction with a broad range of caveats-relevant issues is evidence of the continuing relevance of caveats in military coalition operations. Still, few contributions attempt to clarify the concept with the precision necessary to make studies of the politics of caveats comparable.

Recall that caveats are not about the initial decisions to partake in the coalition or choose what military capabilities to contribute. Analytically speaking, these issues are separate from the decision as to whether, when, how, and where to apply caveats to a national military contingent assigned to a coalition force. Having discussed analytical, historical and operational contexts of caveats in the previous chapters, what remains to do in the present chapter is to narrow down the precise analytical properties and empirical indicators of the concept. Constructing definitions goes beyond the thrills of the analytical exercise. To move the study of

© The Author(s) 2019
G. Fermann, *Coping with Caveats in Coalition Warfare*,
https://doi.org/10.1007/978-3-319-92519-6_4

the politics of caveats forward, we require a concept of caveats that make the phenomenon recognizable, possible to measure and capable of contributing to the construction of reliable data. The conception of national reservations on the use of force also needs to distinguish clearly between different kinds of caveats.

FRAGMENTS OF CAVEATS

In this effort, the literature on the politics of caveats is both suggestive and confusing. How can we systematically compare reports on and studies of coalitions and contributing states on the prevalence of caveats when we have yet to agree upon the conceptual and operational delimitation of the phenomenon? Because studies refer to only partially, overlapping phenomena, are almost unspoken about some important conceptual properties, and are either too broad or too narrow in their conception of caveats to capture the essence or complexity of the phenomenon, any attempt to put together a systematic database on caveats is premature or suggestive at best.

In reviewing the literature on the politics of caveats, we find some studies to refer to caveats in terms of how national reservations have prevented military units to participate in offensive and risky military operations (Mello 2014: 113–114; Ringsmose 2010: 328; Sky 2007: 16). Other contributors are focusing on the controlling function of national staff officers assigned to coalition command to make sure Force Commander uses national contingent in accordance with what the coalition has agreed upon, or on the discretion granted by governments to the most senior member of a nation's contingent to veto orders from the multilateral chain of command—so-called red card-holders (Saideman and Auerswald 2012: 69–70; Høiback 2009: 23–34; Young 2003: 115).

In some studies, researchers conceptually link caveats to national constitutional conditions that lead to reserved coalition-behavior (e.g., Van der Meulen and Kawano 2008; Koschut 2014: 351–354). One study allows caveats to cover the whole range of financial, logistical-, and capacity-related restrictions regarding the military robustness of the contingent (Brophy and Fisera 2010: 1). A final conception of national reservations on the use of force, especially in the context of the NATO campaign in Afghanistan, is the geographical limitations on force mobility states impose on their military contingents (Kay 2013: 109–110; Noetzel and Rid 2009: 75; Noetzel and Schreer 2009: 532).

The inconsistent use of the caveats concept across scholarships implies that it is problematic to compare and lump together the ISAF coalition staff-finding that some 70 instances of caveats were at work in Afghanistan (Bergen 2011: 189) with, say, Otto Trønnes' (2012) finding that the Norwegian government applied tenfold with caveats to its contributions to the multinational war effort in Afghanistan from 2001 through 2008. Also, the nuanced conception of caveats Per Marius Frost-Nielsen (2016: 10–19) applies to research on Danish, Dutch, and Norwegian caveats-use in the 2011 Libya coalition operation (2017: 3–4), deviates considerably from the notion of caveats Ben Lombardi (2008) uses in the empirical mapping of German caveats-policies in Afghanistan.

The lessons to learn from the conceptually fragmented literature is, one, that any empirical research program on the politics of caveats need to establish a reasoned and measurable conception of caveats at the outset that takes into consideration the best of what the caveats-literature has to offer. In some instances, scholars cross the border to an adjacent phenomenon. More often, the confusion and lack of compatibility are due to scholars emphasize different aspects of the phenomenon of caveats. Still, collective inconsistency in how the concept is defined and measured impedes systematic research and renders comparability of research findings in doubt. Second, any generalizing research ambition in the study of the politics of caveats requires the concept of caveats be "liberated from its contextual particularities in time and space to transcend the local experience" (Platt 2007: 120). Finally, the conceptualization should be extensive enough to grasp the complexity of the phenomenon. However, the scope of the concept should also be narrow enough to exclude related, but different phenomena such as the decision to contribute militarily to coalition forces, other instruments of foreign policy-making (Gerring 1999; Fermann 2013: 72–83), other constitutional checks on the use of force abroad (Mello 2014: 6), and other ways of imposing national control on the use of force in coalition operations (Auerswald and Saideman 2014: 5–12), as discussed in Chapter 2.

Defining Properties of Caveats

What, then, are the basic attributes of caveats? We took the first step in the introductory chapter by defining caveats in terms of *national reservations on the use of force in coalition contexts* (Frost-Nielsen 2016: 10–11). In so doing, we distinguished caveats from other adjacent phenomena such as the decision whether to participate in the coalition or the decision to offer a substantial military unit rather than merely a symbolic presence. Caveats thus relate to nation-specific conditions for the use of force in the field—when, how, to what extent, and where within the area of deployment of the coalition.

To further distinguishing caveats from other adjacent phenomena that are likely to have different causes and consequences, the second step is to specify a set of criteria such national reservations on the use of force need to fulfill to qualify as caveats. Crucially, we reserve the concept of caveats for *self-imposed national reservations* on the use of force that are the result of *politically motivated* decision-making. The importance of emphasizing the political nature of caveats is that it captures the reality of caveats as a foreign policy instrument that reflect some national priorities. This political understanding of caveats distinguishes the concept from restrictive military behavior that results from uncoordinated action and incompetence due to chance, or some technical, logistical, managerial or financial limitation (Findlay 2002: 354–359).

In reviewing the literature, we find that some force-behavior that is due to lack of coordination and other kinds of resource-limitations are easily mistaken for caveats. The Norwegian Air Force refrained from participating in offensive actions against Serb forces during the 1999 NATO intervention in Serbia over the Kosovo conflict. This decision was not due to some political motivation, but because the Norwegian Air Force at the time was not capable of executing precision bombing at night. The Norwegians were left to execute purely defensive missions in the airspace above the Adriatic Sea (Anrig 2015: 270). Compare the technical restriction on the Norwegian use of air power in the Kosovo conflict with the Dutch politically motivated restrictions on the use of their F-16 fighter jets in the 2011 intervention in Libya. The Dutch military had the necessary equipment and training to engage in the offensive precision bombing, but the Dutch government decided for political reasons to limit their contribution to the patrolling of the Libyan airspace (Frost-Nielsen 2017).

We further suggest the concept of caveats as national reservations on the use of force be limited to military contingents *subordinated to a unified chain of command*, and to military conduct relating to some *common mechanism for the regulation of the use of force*. This dual specification draws a line against secondary, non-combatant and defensive operational contributions such as the facilitation of military hospitals and other support functions. Only combat units qualify as instruments of warfare, and only combat units are thus potential recipients of reservations on the use of force. More obvious, the above specification of definition also rules out unilateral military operations from the empirical universe of caveats. Scholary discussion of caveats as defined is relevant only in a multinational context. However, the analytical condition that national contingent subordinated to a coalition chain of command also implies an expansion of the empirical boundaries of the caveats-concept. Caveats are not limited to the inclusion of UN, NATO and "Coalition-of–the-Willing" operations after the end of the Cold War. As discussed in Chapters 1 and 3, military history indicates that the application of national reservations on the use of force is as old as coalition warfare. We thus suggest extending the generic scope of the concept of caveats to include all past, present and future coalition forces that fulfil the several criteria argued in the present section. Finally, the above specification of the concept anticipates two essential ingredients in the subsequent arguing of caveats empirical indicators related to the extent coalition command overruled by national representatives in the coalition chain of command, and to national behavioral deviations from the use of force permitted for in coalition rules of engagement (RoE).

A further specification of the definition of caveats is our decision to include not only national reservations that are documented formally in political statements and admitted in particular operational codes of conduct, but also informal, undeclared, and even unadmitted use of caveats that show in *actual behavior*. Behavioral practice-patterns may be demanding to reveal in empirical research, and further emphasizes the need for the precise conceptualization of what caveats are, and how the phenomenon manifest.

An example of informal and undeclared caveats is the German behavior in Afghanistan. Germany placed significant caveats on military action in ISAF—perhaps more so than any other country. Still, German officials tried their best in public statements to conceal their restrictive policies on the use of force to avoid the image of Germany as a risk averse and uncommitted ally (Auerswald and Saideman 2014: 146–147). In such cases, it is the more crucial that rather than relying on official statements, we reveal caveats through the systematic empirical study of contingent's actual behavior as compared to coalition RoE. The deviating, but unstated behavior that cannot be explained as technical or other limitations, may turn out to be politically motivated caveats on the use of force. If declarations of intent are confusing or absent, observe the operational pattern relating to the use of force.

The study of actual behavior is particularly important when the nature of operation changes. The change will potentially affect states' political views on the operations, and, in turn, how they choose to interpret the coalition RoE. NATO's air operations over Libya in 2011 was initially justified by the necessity to prevent civilian atrocities. When this objective was accomplished, the conflict on the ground went into a stalemate that induced key NATO members to turn the originally defensive nature of the mission into the offensive. At this point, several other less enthusiastic coalition members started informally apply restrictive caveats on their use of force (Bouchard 2012: 134).

Finally, we draw on Per Marius Frost-Nielsen's (2016: 15–16) line of reasoning to argue a symmetrical understanding of the term "reservations" to allow for both *restrictive* and *permissive* interpretations of the caveats-phenomenon. The literature indicates that the large majority of national reservations on the use of force are restrictive. Still, the history of caveats shows at least a handful of reservation instances that were of a permissive kind. For instance, the Dutch contingent in ISAF reserved itself the right to use their air support unit even when vetoed by coalition command (Auerswald and Saideman 2014: 166). We may as well classify the regular practice of the United States to insist on the prerogative of having an American general lead the coalition force as an instance of permissive caveats. This final specification of the theoretical definition of caveats represents an extension of the concept. Further support for a symmetrical conception of caveats is the literal meaning of the word "caveat" as a "clause or a warning that embodies specific limitations, conditions, or stipulations" (Concise Oxford English Dictionary 2006: 225). While "limitations" supports the predominant impression that caveats as national reservations on the use of force much more often than not are about restricting the use of force, "conditions" and "stipulations" are inclusive terms open for both restrictive and permissive caveats. A theoretical argument in support of a symmetrical understanding of national reservations on the use of force is that both restrictive and permissive caveats explained as means for the fine-tuning of national self-interest. Whereas restrictive caveats signal reluctant participation, permissive caveats may signal greater geopolitical responsibilities, enthusiastic participation, or low tolerance for own losses.

By reasoning several additional properties of the theoretical definition of caveats, the initial concept of caveats as national reservations on the use of force has evolved into an analytical construct with considerably higher resolving and phenomena-discriminating power:

Caveats are politically motivated, national conditions for the use of force in a coalition force, where military contingents are subordinated to a unified chain of command and relate to some common regulation of the use of force. Particular national conditions for the use of force can be either of a restrictive, or a permissive kind, and may be formally recognized as such, or be informal, undeclared, and even unadmitted by the force-contributing state, only to be observed in actual force-deviating behavior not related to lack of capacity or coordination.

The question remains, however, how are we—more precisely—to measure national reservations on the use of force as restrictive or permissive caveats in actual behavior in the theater of war? We have given some clues in the previous.

OBSERVING CAVEATS: YARDSTICKS OF MEASUREMENT

In arguing how to measure the phenomenon of caveats, we need to relate observable caveats-behavior to the primary mechanisms for the regulation of the use of force in military organizations. Indeed, caveats have no observable meaning if not related to some regulatory framework at the level of the coalition. Hence, we suggest (a) coalition RoE used as a yardstick measuring caveats on how to use force, and (b) coalition/alliance burden-sharing settlements and bilateral force-agreements between coalition leader and contributing states to reveal caveats on task assignment and geographical mobility. Recall from the previous chapter that the conception of RoE includes two dimensions. The first dimension relates the RoE guidelines to the conditions need fulfilled to use force in the first place, and the enforcement actions allowed under prevailing circumstances. This particular dimension of RoE clarifies, to various degrees of specificity, when, where, against whom, and how military force may be used. The second dimension relates to how delegating the RoE is as to who has the authority to make judgments about conditions for using force and approve of specific military actions at different levels of command. How may we use this two-dimensional understanding of RoE to operationalize the concept of caveats for empirical research?

It is precisely because the force-regulating guidelines of RoE inevitably represent some political priorities this mechanism qualifies as a benchmark against which national deviations on the use of force be measured. Even if RoE vary in robustness and specificity across coalitions, there always be a RoE against which national use of force be measured. This common denominator makes national deviations from coalition RoE a context-independent yardstick capable of supporting generic research ambitions (Adcock and Collier 2001). A government applies some sorts of caveats to the extent the national military contingent in its operational practice and for political reasons *deviates* from the coalition RoE, whether it be in (i) the *conditions* for the use of force and the *kind of force* permitted, or (ii) in the questioning of who has the *authority* to make judgements on the use of force in the coalition chain of command. By juxtaposing coalition RoE against national behavior (statements and actions), we recognize caveats in national deviations from the *force-regulating guidelines* of the coalition RoE, and in an assertive *government interfering in the coalition chain of command* utilizing military personnel as operational agents of the government. National interference in coalition chain of command is observable in the discretion granted by governments to the most senior member of a nation's contingent to veto orders from coalition chain of command ("red cardholder"). "Red carding" is institutionalized in NATO and beyond, and implies that national commanders "can choose not to obey orders coming from the multinational chain of command if the [national] commander views the orders as being illegal, contrary to his or her country's national interest, or excessively reckless" (Auerswald and Saideman 2014: 5). National interference to veto (or to instigate) particular use of contingent is also observed in the more subtle intervention of national staff officers assigned to coalition command for coalition planning purposes but also to ensure that Force Commander uses national contingent by what coalition has agreed to. Crucial, RoE regulates the use of force at multiple levels of operational command, from Force Commander and contingent commander down to the private soldier. This implies that national reservations on the use of force also observed on the tactical level of the platoon commander and on the individual level of the fighter pilot and gun operator as deviations as to who has the authority to make decisions on conditions for use of force and how. Which, in turn, implies that that delimiting research to only cover those officers assigned official "red carding" authority will leave much caveats-relevant data untouched.

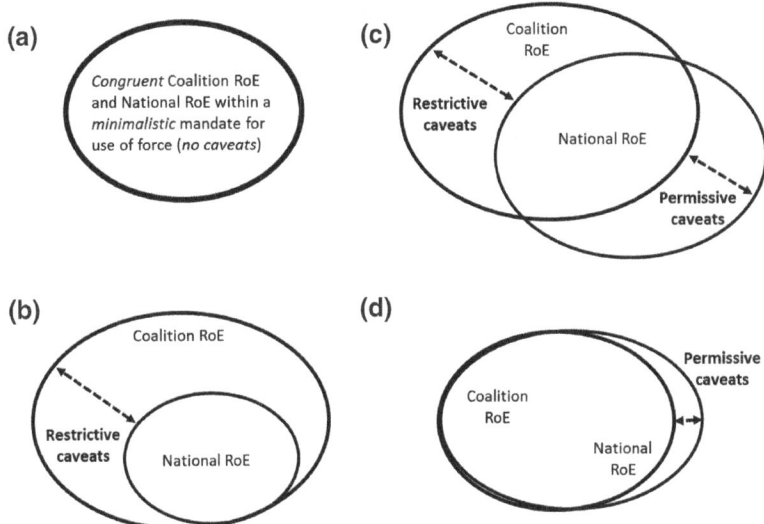

Fig. 4.1 Four constellations of caveats (elaboration on Husby 2015: 23)

Figure 4.1 illustrates how national interpretations and practicing of coalition RoE ("national RoE") may deviate (or not) from the letter and intent of the coalition RoE, and thus register as restrictive or permissive caveats in four hypothetical examples along either or both of the two dimensions of RoE discussed so far (regulation of use of force; national interference in coalition chain of command). The different size of the ellipses indicates the degree of robustness of the RoE (national or coalition) as indicated in the previous chapter. The difference in size between the ellipses of coalition RoE and national RoE is there to indicate the scope of divergence in robustness between the two. The lack of complete overlap between national RoE and coalition RoE constitutes caveats (national reservations on the use of force) and is, in principle, possible to measure empirically.

Case "A" is the special case where the ellipses of national RoE and coalition RoE are completely overlapping. The smaller, merged ellipses indicate a coalition RoE that is not very robust, and we have to do with a national contingent that does not apply any reservations whatsoever.

Case "B" describes a rather robust coalition RoE and a national contingent applying ample restrictions on the use of force. The government's use of restrictive caveats may have been provoked by the extensive use of force allowed for in the coalition RoE. Or possibly because the troop-contributing government in question is a minority coalition government which is hard-pressed to convince all cabinet members that participating in the coalition force is in the best interest of the country. Or perhaps a combination of the two.

Case "C" is more complex, and possibly also the most unlikely scenario. Here both the coalition RoE and the national RoE allow for the extensive use of force (robust RoEs). However, the large mismatch between the two RoE is likely to create tensions within the coalition. Especially since the caveats in question are of both a restrictive and permissive kind.

Finally, case "D" describes a situation where both the coalition RoE and the national RoE are rather moderate concerning the use of force, and only slightly diverge. The most interesting aspect of this particular case is the permissive caveats applied by the national contingent in question. Emboldened by the moderately robust mandate, and urged by the need to confirm solidarity with alliance partners, perhaps this was the right occasion for a risk aversive alliance partner to prove itself? Perhaps the extra effort in the use of force was applied due to some particular interests in the area of operation? These are, of course, pure speculations, but point toward the explanatory ambition of the empirical research program in subsequent chapters.

Finally, we may observe national reservations on the use of force in the *extent to which coalition is delegated authority to make full use of the operational capacity of the national contingent.* When a national military unit is assigned to coalition command, and it is part of the a political settlement that the contingent is to be deployed in a particular area, or that the unit is assigned a specific role which is functionally limited to the execution of particular tasks, these conditions are not adequately registered as deviations from the coalition RoE in either of the two dimensions discussed. This is because, in military and operational context, the RoE is not the mechanism regulating what missions and tasks contingents asked to execute, and where to deploy. Such reservations on the use of force are instead regulated in initial burden-sharing and force-generating settlements in NATO, or in bilateral force-agreements negotiated between the coalition-leading state and contributing government. To preemptively decline to take on tasks assigned by Force Commander on the basis of such political settlements still be considered the result of a deliberate political decision to not fully subordinate the national contingent to coalition command.

This empirical indicator is context-specific and thus requires knowledge about the particular states' military capabilities to be able to judge whether the geographical and task-specific limitations are due to political reservations, or due to some military, technical or financial limitations.

Table 4.1 Typology of caveats—identifying and classifying national reservations on the use of force

Operational dimensions of caveats		
National deviations from the force-regulating coalition rules of engagement (RoE) in terms of when use of force is permitted and how on any level of operational command	National interference in coalition chain of command in discretion to veto orders by means of designated red card-holder, staff officers assigned to coalition command, or personnel further down coalition chain of command	Particular national conditions on the extent to which coalition is delegated the authority to make full use of the operational capacity of the national contingent as to where, when, and how contingent be deployed and used in theater of war
An instance of caveats (restrictive or permissive; officially recognized or not)	An instance of caveats (restrictive or permissive; officially recognized or not)	An instance of caveats (restrictive or permissive; officially recognized or not)

We have seen several instances how caveats justified in political settlements play out in the field. In NATOs operations in Afghanistan, it was a continuing operational problem that troop-contributing states did not allow NATO to regroup military units across geographical sectors of operation based on where needed the most. The units in question did operate according to coalition RoE. However, the contributing governments put severe restrictions on the mobility of forces, and thus what tasks able to execute (e.g., Trønnes 2012: 68–71).

Table 4.1 summarizes the operational definitions coming out from the conceptual discussion on the empirical footprint of caveats as national reservations on the use of force. Following several lines of reasoning, we suggest that the concept of caveats is observable as (i) national deviations from the coalition RoE in terms of when use of force is permitted and how; (ii) in national interference in the coalition's chain of command, and (iii) in particular national conditions as to where, when and how the national contingent be deployed and used in theater of war.

Arguably, some of the conceptual fragmentation in the literature originates from the fact that in many studies only one of the several operational dimensions are applied, and in some cases inconsistently so. By systematically applying the more complex battery of measurement indicators, we may observe and reflect upon previously undetected instances and kinds of caveats. Common to all the three operational dimensions is the fundamental attribute that national reservations on the use of force are not the reflection of some lack of military capacity, insufficient coordination or chance, but rather the result of a calculated politi-cal decision, serving some foreign policy-purpose. However, in classify-ing particular caveats, it is, depending on the research question, crucial to consult also other distinguishing properties of the concept relating to whether the caveats in question are of a restrictive or permissive kind, and the extent to which caveats used are officially recognized and admitted.

A strong indication of the usefulness of the conceptual contribution of the study is the capacity to distinguish between different classes and types of caveats. As defined, the conception of caveats as national reservations on the use of force in coalition forces comes in many shapes. Restrictive caveats show in the unwillingness of national contingents to fight when they are expected to according to the common coalition RoE. It shows in the sometimes, informal practice of having a national representative—a red card-holder—acting as a gate-keeper or veto-player in the coalition chain of command for missions and tasks the government does not want to take on. Restrictive caveats also show in the unwillingness to fight and contribute outside the assigned area of deployment, or at night, even if the operational threat-situation so requires and the Force Commander so orders. Restrictive caveats furthermore show in a defensive posture when offensive action is required and allowed for in coalition RoE. Permissive caveats, on the other hand, appears in the alliance leader's insistence to take on the military leadership of the coalition, and in a participating state's willingness to use excessive force when contingent is in dire straits.

The proposal for a multi-dimensional operationalization of national reservations on the use of force invites research explaining why coa-lition members may choose to apply particular kinds of caveats, offi-cially admitted or not. The main challenge for empirical research is, of course, to document that informal or officially unrecognized cave-ats are at work and to render probable that the reservations in ques-tion are politically calculated. Methodological challenges and remedial strategies related to the gathering of data is discussed in Chapter 11.

REFERENCES

Adcock, R., & Collier, D. (2001). Measurement Validity: A Shared Standard for Qualitative and Quantitative Research. *American Political Science Review*, *95*(3), 529–546.

Anrig, C. F. (2015). The Belgian, Danish, Norwegian, and Dutch Experiences. In K. P. Mueller (Ed.), *Precision and Purpose: Airpower in the Libyan Civil War*. Santa Monica, CA: RAND.

Auerswald, D. P., & Saideman, S. M. (2014). *NATO in Afghanistan: Fighting Together, Fighting Alone*. Princeton, NJ: Princeton University Press.

Bergen, P. L. (2011). *The Longest War*. New York, NY: Free Press.

Bouchard, C. (2012). Lessons Learned from Operation Unified Protector—A Commander's Perspective. *Papers of the Royal Norwegian Air Force Academy, 27*, 127–137.

Brophy, J., & Fisera, M. (2010). *National Caveats and Its Impact on the Army of the Czech Republic*. http://user.unob.cz/fisera/files/clanky/National_Caveats_Short_Version_version_V_29JULY.pdf.

Concise Oxford English Dictionary. (2006). *Concise Oxford English Dictionary*. Oxford: Oxford University Press.

de Nevers, R. (2007). NATO's International Role in the Terrorist Era. *International Security, 31*(4), 34–66.

Fermann, G., ed. (2013). *Utenrikspolitikk og norsk krisehåndtering*. Oslo: Cappelen Damm Akademika. https://www.cappelendamm.no/_utenrikspolitikk-og-norsk-kriseh%C3%A5ndtering-gunnar-fermann-9788202378691.

Findlay, T. (2002). *The Use of Force in UN Peace Operations*. Oxford: Oxford University Press.

Frost-Nielsen, P. M. (2016). *Betingede forpliktelser. Nasjonale reservasjoner i militære koalisjonsoperasjoner*. Ph.D. Dissertation in Political Science, Department of Sociology and Political Science, Norwegian University of Science and Technology (NTNU), Trondheim.

Frost-Nielsen, P. M. (2017). Conditional Commitments: Why States Use Caveats to Reserve Their Efforts in Military Coalition Operations. *Contemporary Security Policy, 38*(3), 371–397.

Gerring, J. (1999). What Makes a Concept Good? A Critical Framework for Understanding Concept Formation in the Social Sciences. *Polity, 31*(3), 357–393.

Høiback, H. (2009). The Noble Art of Constructive Ambiguity. *Oslo Files on Defence and Security, 3*, 19–39.

Husby, G. (2015). *Fra hull i luften, til hull i Gaddafis bunker. Bruk av politiske reservasjoner på norsk militærmakt i flernasjonale koalisjonsoperasjoner. En komparativ studie av F-16 bidragene i Kosovo, Afghanistan og Libya*. Master Thesis in Political Science, Department of Sociology and Political Science. Norwegian University of Science and Technology (NTNU), Trondheim.

Kay, S. (2013). No More Free-Riding: The Political Economy of Military Power and the Transatlantic Relationship. In J. H. Matlary & Magnus Petersson (Eds.), *NATO's European Allies—Military Capability and Political Will* (pp. 97–120). Hampshire: Palgrave Macmillan.

Koschut, S. (2014). Transatlantic Conflict Management Inside-out: The Impact of Domestic Norms on Regional Security Practices. *Cambridge Review of International Affairs, 27*(2), 339–361.

Kreps, S. (2008). When Does the Mission Determine the Coalition? The Logic of Multilateral Intervention and the Case of Afghanistan. *Security Studies, 17*(3), 531–567.

Lombardi, B. (2008). All Politics Is Local: Germany, the Bundeswehr, and Afghanistan. *International Journal, 63*(3), 587–605.

Marten, K. (2007). State-Building and Force: The Proper Role of Foreign Militaries. *Journal of Intervention and Statebuilding, 1*(2), 231–247.

Mello, P. A. (2014). *Democratic Participation in Armed Conflict.* Houndmills, Basingstoke: Palgrave Macmillan.

Noetzel, T., & Rid, T. (2009). Germany's Options in Afghanistan. *Survival, 51*(5), 71–90.

Noetzel, T., & Schreer, B. (2009). Does a Multi-tier NATO Matter? The Atlantic Alliance and the Process of Strategic Change. *International Affairs, 85*(2), 211–226.

Platt, J. (2007). Case Study. In W. Outhwaite & S. P. Turner (Eds.), *The Sage Handbook of Social Science Methodology* (pp. 102–127). London: Sage.

Ringsmose, J. (2010). NATO Burden-Sharing Redux: Continuity and Change After the Cold War. *Contemporary Security Policy, 31*(2), 319–338.

Saideman, S. M., & Auerswald, D. P. (2012). Comparing Caveats: Understanding the Sources of National Restrictions upon NATO's Mission in Afghanistan. *International Studies Quarterly, 56*(1), 67–84.

Sky, E. (2007). Increasing ISAF's Impact on Stability in Afghanistan. *Defense and Security Analysis, 23*(1), 7–25.

Trønnes, O. (2012) *Mapping and Explaining Norwegian Caveats in Afghanistan from 2001 to 2008.* Master thesis in Political Science, Trondheim, Department of Sociology and Political Science, Norwegian University of Science and Technology (NTNU).

Van der Meulen, J., & Kawano, H. (2008). Accidental Neighbours: Japanese and Dutch Troops in Iraq. In J. Soeters & P. Manigart (Eds.), *Military Cooperation in Multinational Peace Operations: Managing Cultural Diversity and Crisis Response* (pp. 166–179). Oxon: Routledge.

Young, T.-D. (2003). The Revolution in Military Affairs and Coalition Operations: Problem Areas and Solutions. *Defense and Security Analysis, 19*(2), 111–130.

PART III

Approaching Caveats

The Epistemological Function of Foreign Policy Analysis in the Empirical Research Program

The purpose of an empirical research program is to facilitate the construction of systematic, reliable, and generic knowledge within a specific field of inquiry. We argued in the Introductory chapter that empirical research programs engage in the formulation of substantive research problems, are grounded in analytical frameworks of interpretation and explanation, and gives direction on methods to gather data and make empirical patterns comprehensible.

Despite some excellent studies on the politics of caveats, the study of caveats still is a nascent field of research. We do not know precisely what conditions, mechanisms, dilemmas, trade-offs, and political concerns that make governments reach for caveats rather than completely withdraw from the coalition, or contribute without any political restrictions on the use of force whatsoever. There is a crude pattern of variation here. However, the empirical patterns are not comprehensively documented. Also, the complex of causes and mechanisms is far from understood.

We are not without clues though. We do know that national reservation on the use of force is fairly common in coalition operations. We acknowledge the practice is hurting the fighting capacity of the force and may undermine the political mandate. Also, it is safe to assume that caveats are applied by governments to balance or safeguard some national interest. Also, there is the indication that some configuration of government tends to use caveats more than other.

In the previous chapter, we reasoned and prepared the concept of caveats for further empirical research. The precise conceptualization

© The Author(s) 2019   71
G. Fermann, *Coping with Caveats in Coalition Warfare*,
https://doi.org/10.1007/978-3-319-92519-6_5

of the phenomenon under scrutiny in the empirical research program was the first main step of the study. The next step in the elaboration of the research program is to argue the analytical framework and the explanatory ambition. This analytical, theoretical, and deductive effort runs through six chapters (and two parts), and starts in the foundational discussion of what constitutes the ontological "hard core" of the program, and ends eventually in the formulation of empirical propositions as to how caveats patterns may be explained.

In the present chapter, we thus write the empirical research program into what is arguably the greatest of all scientific puzzles. That is the epistemological debate on how science moves forward, and how research communities may arrive at new knowledge and insights. We liberally make use of Science Philosopher Imre Lakatos' conception of the scientific research program (1978) as a heuristic device to reason the epistemological function of the approach of Foreign Policy Analysis (FPA), which is foundational for the analytical framework of the program. A main take-home message from the discussion is that the function of the FPA approach is directional rather than a point of departure for deducing empirical propositions. Another conclusion is that we assess the approach of FPA on its fruitfulness in inspiring research questions and supervising the choice of middle-range theory. However, as the "hard core" of the analytical framework of the research program, the FPA approach is not by itself equipped to be exposed to any theory testing refutation procedure.

RESEARCH PROGRAMS IN THE STUDY OF THE NATURE OF KNOWLEDGE

In the study of the nature of knowledge (epistemology), the conception of research program has been attributed a particular meaning. It was Scientific Philosopher Imre Lakatos, who first argued the evolution of knowledge in terms of the scientific research program (1978). We may gain reflective traction by framing our approach to the politics of caveats within Lakatos' conception of a scientific research program (Caldwell 1991).

First, it is useful to put Lakatos' contribution in context. Part of Lakatos' work was an effort to bridge Karl Popper's view of scientific evolution as one in which theories are strengthened or rejected according to the hypothetical-deductive logic of falsification (Popper

1963), with Thomas Kuhn's sociological approach to the evolution of knowledge (Kuhn 1962).

Thomas Kuhn argues that scientific practices do not nearly fall in line with Popper's ideal requirements in the hypothetical-deductive logic of falsification. Instead, Kuhn's sociological approach supports the notion that actual research behavior includes abrupt paradigm shifts that are not well accounted for by Popper's ideal model for the refutation of bad science.

Paul Feyerabend takes Kuhn's notion of radical shifts in scientific paradigms to the extreme by arguing that the practicing of scientific inquiry fully depends on historical context. He resists the notion that a particular method for refutation and validation shall discipline scientific inquiry. Instead, Feyerabend argues that the study of scientific practices shows that "everything goes," and submits an invitation to epistemological anarchy to replace rationalism in the theory of knowledge (1975).

Not surprisingly, the epistemological "anarchism" of Feyerabend does not provide any concrete guidance to the intellectual effort of constructing an empirical research program on the politics of caveats, since he rejects the validity of any particular guidelines ("anything goes"). As to the logical "idealism" of Popper, this approach to cumulative science would seem to be tailor-made for the natural sciences where law-like relationships are discovered, and where test methods exist that allow for the more clear-cut refutation of hypotheses. Concerning Kuhn's sociological "revolutionism," our more limited project hardly fills the boots of any radical paradigm shift since our effort is more about applying existing knowledge and analytical frameworks for a specific research purpose. We consider FPA a complementary and integrative approach, not as a competing approach to International Relations (IR) and Comparative Politics (CP) approaches.

What remains to consider then is Lakatos' conception of the scientific research program. Although developed with the natural sciences in mind, Lakatos' approach to how new knowledge is constructed fills the "pragmatic" middle ground between Popper's logical "idealism" and Kuhn's sociological "revolutionism." Lakatos' work is useful for our purposes for reasons which will become evident. For now, accept that to argue a Political Science empirical research program on the back of a Lakatosian theory of knowledge is to demonstrate how the insights from the study of epistemology may be translated and applied for different research purposes.

THE LAKATOSIAN NOTION OF THE SCIENTIFIC RESEARCH PROGRAM

A Lakatosian research program is founded, first, on a "hard core" of the-oretical assumptions that cannot be abandoned or altered without aban-doning the program altogether. Hence, we need to argue what is the "hard core" (some approach) of our framework for analysis. This core ontology of the research program is considered credible, robust, and axi-omatic enough to be excepted from further empirical questioning and refutation attempts. What needs to be argued, however, is the poten-tial and continuing fruitfulness of the "hard core" to inspire promising research.

Second, Lakatos requires us to reason more specific theories, so-called "auxiliary hypotheses." We require of such substantial "middle-range" theories not only that they contribute to explaining the relevant phe-nomenon in question, but also that they conform with and thus protect the "hard core" of ontological assumptions underpinning and at the heart of the research program. Note that theories are simplified ontolo-gies of reality. Political theories explain how the political world works by emphasizing certain aspects and leaving out others. Theories provide a causal story based on the particular ontological assumptions of the the-ory in question (Mearsheimer and Walt 2013: 431–432).

Third, Lakatos argues that the "protective belt of auxiliary hypothe-ses" (middle-range theories) are expendable and should be kept only as long as the theories can harmonize relevant facts on the ground (the data) with the fundamental assumptions of the "hard core" of the research program. While the "hard core" of the research program is taken for granted (until it is left for some other, more productive research program), the "auxiliary hypotheses" must stand the test against empirical data on a much more daily basis.

Finally, a progressive research program is one in which the "protective belt of auxiliary hypotheses" can explain new facts, or existing facts better, without conflicting with the "hard core." Degenerative research programs, on the other hand, are holding on to "auxiliary hypotheses" consistent with the "hard core" even after the substantial theories in question have been refuted in the confrontation with data (Lakatos 1978: 47–51).

The Lakatosian research program provides a framework within which research is conducted on the back of axiomatic first principles (the "hard core") shared by those involved in the research program, and accepted as such without further proof or debate (Caldwell 1991). What still is open for debate are what research questions to be investigated; how concepts should be defined and operationalized; what specific middle-range theories are the most promising to apply; what hypotheses be deduced; what case selections to be made; what concessions should be granted to parsimony; and what research methods to make use of.

What, then, is the "hard core" of the empirical research program on the politics of caveats we are about to propose? The approach to the analytical framework of the empirical research program rests on seven interrelated assumptions, which go far deeper into the interplay between global and domestic politics than the more top-down approach of IR and the unit crosscutting approach of CP.

First, the primary agency in the political study of caveats rests with the foreign policy-making state.

Second, the phenomenon of "caveats" is acknowledged as a potentially useful foreign policy instrument of the state, rather than merrily a challenge to "the effective usability of troops" (Jones 2004).

Third, the state is not a unitary actor, but a social organization made up of decision-making and implementing institutions and actors, that are sensitive toward external environments and political principals.

Fourth, the way the foreign policy-making process is institutionalized and embedded in collective self-conceptions, culture, and norms influence policy decisions on foreign policy preferences, means, and behavior.

Fifth, the state is politically independent (sovereign), but its freedom of action in international affairs is limited, depending on relative strength, political skills, and issue-area.

Sixth, the state's foreign policy scope for manoeuvering is influenced by structures, actors, interests, and sentiments in both the global and domestic environments.

Finally, key decision-makers' perceptions and narratives of the dual environments strongly influence what is deemed politically feasible to do in foreign policy-making, including decisions related to alliance dynamics and participation in coalition forces. Still, there is an opening for political engineering in foreign policy-making. Key decision-makers may engage in political, creative action in a calculated manner, with the aim of exploiting the scope for manoeuvering offered by structural opportunities, and the actions or passivity of other actors.

THE APPROACH OF FOREIGN POLICY ANALYSIS AS THE "HARD CORE" OF THE RESEARCH PROGRAM

What approach to politics fits the above prescription for an epistemological "hard core"? We are aware of only one approach to the study of politics which satisfies the state-centered, bottom-up approach assumptions outlined above, and which require scholars to look for explanations of caveats on multiple levels of analysis, as well as to engage in the meticulous empirical study of actual foreign policy decision-making and implementing processes. This approach is the eclectic framework of FPA (Gerner 1992).

Unlike most approaches to IR, which are pre-occupied with making sense of international and transnational patterns of interaction (conflict, peace, cooperation), the bottom-up approach of FPA is tailor-made to invite explanations of foreign policy outcomes such as why do states find it in their interest to enter into trade agreements, spend X% of GDP on foreign aid, decide to participate in alliances, justify their foreign policies as they do, or decide to apply caveats on their military contributions to coalition forces.

As will be substantiated in the subsequent chapters, it follows from the multi-level and decision-making emphasizes of FPA that the framework gives priority to the epistemological criteria of explanatory power and ontological realism over parsimony (Fermann 2013; Neack 2013; Hill 2003). Arguably, this might be precisely what we need in a nascent research field such as ours, where little should be taken for granted, and much checked against empirical data at close distance to actual decision-making processes.

This leads us to the next question; what does the "protective belt of auxiliary hypotheses" looks like in an empirical research program grounded in the FPA approach? Recall that the theories accepted in the "protective belt" are required, first, not to conflict with the "hard core"-assumptions of FPA. Second, the theories in question should be able to explain at least some variation in the dependent variable (say, the use of caveats in foreign policy). The second requirement is an empirical question, implying that it depends on the results of future empirical research on caveats based on the direction of key FPA-assumptions (the "hard core"). As to the first requirement, there is a multitude of social and political theories qualifying for potential use within the eclectic FPA approach (see Chapters 7–10).

Due to the multi-level and decision-making emphasizes of FPA, the "protective belt" of theories around the "hard core" of the proposed empirical research program is potentially wast. Since its inception in the early 1950s, FPA has drawn upon intellectual imports from all sub-branches of Political Science and beyond: Insights from IR related to anarchy, power, security, norms, identity, interdependence, and transnational division of labor are used in FPA to explain how the global environment may influence foreign policy-making (Fermann 2013: 103–108).

Likewise, FPA borrows insights from the subfields of Political Behavior and Political Communication in shedding light on how domestic factors may influence foreign policy-making and on how foreign policies are communicated and justified. Furthermore, Public Policy and Administration, and organizational theory are indispensable in explaining how institutions structure, and thus influence the foreign policy-making process; as is Political Psychology in clarifying the conditions for individual decision-makers making a difference for the outcome of foreign policy-making processes (Fermann 2013: 108–119; Hudson 2005).

The reason why FPA can make legitimate claims to such various bodies of literature is partly that foreign policy-making is studied as a complex, social and political process, and not "assumed away" in the proverbial "black box" of the unitary state. Even more, because foreign policy-making is about balancing domestic interests and forces against global threats, risks, and possibilities. The fact that foreign policy-making is going on at the interface between global and domestic politics requires the student of FPA to seek out approaches theorizing both the domestic and global environments of the state, as well as the institutions and people responsible for making foreign policy decisions.

However, to be potentially useful within an FPA approach, bodies of Political Science middle-range theory on different levels of analysis require translating for foreign policy purposes. Translation is necessary because most approaches to the study of politics are not tailor-made to illuminate foreign policy phenomena. To illustrate: IR theory is in FPA not used to explain international phenomena such as interstate war, peace, and international cooperation, but instead applied to explain states' foreign policy goals, means, and behavior toward the global political environment.

How do we justify the adaptation of IR theory for foreign policy-explaining purposes in light of, e.g., Kenneth Waltz' (1996) emphasizing that "international politics is not foreign policy" (1996)? Mainly by assuming, or empirically demonstrating that key decision-makers' perceptions and decision-making cultures reflect the global political environment as reasoned in particular IR theories. As mentioned, theories on politics are ontologies, sets of assumptions on how politics works. To the extent decision-makers share these ontologies/narratives, their decision-making will be influenced by these perceptions. For instance, it is entirely plausible that insights from, say, Political Realism apply to FPA in explaining why foreign policy decision-makers may choose to bring nations into alliances, and why states apply reservations on the use of force when participating in coalition operations.

Likewise, theories from the field of Political Behavior on how societal interests and sentiments are communicated and brought to bear on the political decision-making institutions are hard to ignore if we are to understand better the implication of the notion that "all politics is local" for foreign policy-making (Moravcsik 1997). The most well known and perhaps the best-developed translation of theories for FPA is Graham Allison's fruitful application of organizational theory. He demonstrates how different models of the decision-making process contribute to explain US policies during the 1962 Cuban Missile Crisis based on data on the crisis management in the US government (Allison and Zelikow 1999).

The primary concern guiding our choice of expendable middle-range theories within the "protective belt of auxiliary hypotheses" is that more than one level of analysis is represented by at least one cluster of theory—say, on global politics, domestic politics, governmental politics, or the politics of implementation. This allows us to account also for the likely interaction between different levels of analysis that plays out in most critical foreign policy decision-making processes (Putnam 1988; Tsebelis 1991). Indeed, the overarching idea of the study is that caveats as a foreign policy instrument is applied by strategic decision-makers at the national level to balance different concerns in alliance politics, domestic politics, and in the politics of controlling the sword in the theater of war.

By now, it has become clear that no FPA of the politics of caveats can be theoretically exhaustive given the huge selection of substantial theories potentially available in Social Science. It is the more crucial that the choice of "auxiliary hypotheses" is well argued and translated for foreign policy research. If the theories applied in the "protective belt" surrounding/protecting the core assumptions of FPA do not deliver on the promise, this is an invitation to seek out other theories from which we can deduce plausible empirical propositions. This is for the sake of learning more about the puzzle, the politics of national reservations on the use of force in multinational military operations.

On the Objection That Foreign Policy Analysis Is Merely a "Pre-theory"

A final issue to address before we embark on a review of what bodies of theory may be engaged in FPA relates to James Rosenau's observation that the approach of FPA is a "pre-theory" rather than a substantial theory capable of standing on its own feet (1966). Based on our previous line of argument, we cannot but agree with Rosenau's assessment. However, we fail to see why this limitation should preclude FPA from making a crucial contribution to the analytical framework of the empirical research program.

Recall that the fundamental reason why the "hard core" of FPA is to be considered an approach rather than a fully fledged theory is because FPA lacks the reasoned ontological content (precise assumptions about system structure, agency, and relationships between actors) that is required of analytical constructs (theory) from which we intend to develop empirical propositions (hypotheses). In particular, "pre-theories" are under-specified in terms of identifying the particular causal factors at work, and in the reasoning of the mechanisms involved in the explanation of relationships. As will be demonstrated in Chapters 7–10, this is precisely why FPA does not pretend to qualify as a fully fledged theory, and explicitly invites the assistance of substantial middle-range theories ("auxiliary hypotheses") at multiple levels of analyses to deliver on its promise to the empirical research program.

The ontological limitation of FPA, the approach shares with other analytical approaches and frameworks developed for the study of politics. For instance, the argument has been made that Johan Galtung's Structural Theory of Imperialism (1971) shares the properties of a "taxonomy" rather than the attributes of a fully developed theory from which hypotheses can be deduced (Van den Bergh 1972). Furthermore, Leonard Schoppa (1993) argues that Robert Putnam's two-level game theory on the study of international negotiations (1988) be considered a "metaphor" rather than a substantial theory because the approach needs to be supplemented ontologically to support the deduction of empirical propositions. Indeed, Putnam admits to this limitation himself (1988: 435).

Even if "approaches," "taxonomies," and "metaphors" do not qualify as proper theories from which testable hypotheses be directly deduced, such "pre-theories" still constitute cohesive, epistemological kick-off planks for the further theorizing of the phenomenon in question. In Lakatosian terms, "pre-theories" such as FPA, Structural Imperialism, and the approach of the two-level games constitute the "hard core" of empirical research programs. We consider such approaches to provide valuable inspiration and direction, and necessary identity and discipline to the further development of their respective fields of research (Black 1962: 242).

REFERENCES

Allison, G., & Zelikow, P. (1999). *Essence of Decision: Explaining the Cuban Missile Crisis*. New York, NY: Longman.

Black, M. (1962). *Models and Metaphors*. Ithaca, NY: Cornell University.

Caldwell, B. J. (1991). The Methodology of Scientific Research Programs in Economics. Criticisms and Conjectures. In G. K. Shaw (Ed.), *Economics, Culture, and Education. Essays in Honour of Mark Blaug* (pp. 95–107). London: Edward Elgar.

Fermann, G. (Ed.). (2013). *Utenrikspolitikk og norsk krisehåndtering*. Oslo: Cappelen Damm Akademika.

Feyerabend, P. (1975). *Against Method: Outline of an Anarchist Theory of Knowledge*. Westwood, MA: New Left Books.

Galtung, J. (1971). A Structural Theory of Imperialism. *Journal of Peace Research, 8*(2), 81–117.

Gerner, D. (1992). Foreign Policy Analysis. Exhilarating Eclecticism, Intriguing Enigmas. *International Studies Notes, 18*(4), 4–19.

Hill, C. (2003). *The Changing Politics of Foreign Policy*. London: Palgrave Macmillan.

Hudson, V. M. (2005). Foreign Policy Analysis. Actor-Specific Theory and the Ground of International Relations. *Foreign Policy Analysis, 1*(1), 1–30.

Jones, J. L. (2004, May 7–10). *Prague to Istanbul: Ambition Versus Reality. Global Security: A Broader Concept for the 21st Century*. Center for Strategic Decision Research 21st International Workshop on Global Security, Berlin. http://csdr.org/2004book/Gen_Jones.htm.

Kuhn, T. S. (1962). *The Structure of Scientific Revolutions*. Chicago: Chicago University Press.

Lakatos, I. (1978). *The Methodology of Scientific Research Program*. Cambridge, UK: Cambridge University Press.

Mearsheimer, J. J., & Walt, S. M. (2013). Leaving Theory Behind: Why Simplistic Hypotheses Testing Is Bad for International Relations. *European Journal of International Relations, 19*(3), 427–457.

Moravcsik, A. (1997). Taking Preferences Seriously. A Liberal Theory of International Politics. *International Organization, 51*(4): 513–553.

Neack, L. (2013). *The New Foreign Policy. Complex Interactions, Competing Interests.* Boulder, CO: Rowman & Littlefield.

Popper, K. (2008 [1963]). *Conjectures and Refutations: The Growth of Scientific Knowledge.* New York: Routledge.

Putnam, R. (1988). Diplomacy and Domestic Politics. The Logic of Two-Level Games. *International Organization, 42*(4): 427–460.

Rosenau, J. M. (1966). Pre-theories and Theories of Foreign Policy. In R. B. Farrell (Ed.), *Approaches to Comparative and International Politics* (pp. 27–92). Evanston: North-Western University Press.

Schoppa, L. J. (1993). Two-Level Games and Bargaining Outcomes. *International Organization, 47*(3), 353–386.

Tsebelis, G. (1991). *Nested Games. Rational Choice in Comparative Politics.* Berkeley: University of California Press.

Van den Bergh, G. v. B. (1972). Theory or Taxonomy? *Journal of Peace Research, 9*(1), 77–85.

Waltz, K. N. (1996). International Politics Is Not Foreign Policy. *Security Studies, 6*(1), 54–57.

CHAPTER 6

The Essence of Foreign Policy Analysis (I): Modeling the Foreign Policy-Making and Implementing Processes

In a guest lecture at our university, the Norwegian Minister of Foreign Affairs was asked: "what principles are applied when the strategic leadership of the Ministry reviews new foreign policy initiatives or contemplate how to respond to new developments?" The minister replied that if they conclude that "Norway could make a difference," the next step would be to determine what "is in Norwegian interests to do" (Fermann 2013: 89).

This response is as if lifted straight out of a textbook in Foreign Policy Analysis (FPA). The reply confirms that both the foreign policy-maker and the FPA scholar look for intellectual support in both opportunity and interest when assessing the choice and implementation of foreign policies (see Neack 2013; Fermann 2013; Carlsnaes 2002, 2004, 2008; Hudson 2007; White 2004; Hill 2003; Webber and Smith 2002; Kubálková 2001; Clarke 1996; Light 1994; Clarke and White 1989). A leading journal in the field of FPA nails the core of the approach as follows:

> Foreign policy analysis is the study of the process, effects, causes or outputs of foreign policy decision-making in either a comparative or a case-specific manner. (Foreign Policy Analysis 2013)

The two main tenets of FPA is (a) the study of foreign policy-making and implementing processes, and (b) the appetite to draw upon bodies of theory at multiple levels of analyses. To conduct FPA is—in part or full—to engage Political Science literature and relevant data to shed light

© The Author(s) 2019
G. Fermann, *Coping with Caveats in Coalition Warfare*,
https://doi.org/10.1007/978-3-319-92519-6_6

on how (i) the global and domestic environments of foreign policy-making state, (ii) the institutionalization of foreign policy-making processes, and (iii) attributes of individual decision-makers may influence (iv) perceptions as to what scope is for political maneuvering, (v) choice of foreign policy preferences and goals, (vi) choice of strategy and calibration of policy instruments (caveats included), (vii) material implementation of policies, and (viii) particular speech acts executed to justify foreign policies toward constituencies and target groups capable of influencing the political costs of policy implementation.

POLITICAL ENGINEERING IN FOREIGN AFFAIRS AND THE FOREIGN POLITICS OF CAVEATS

This is the FPA approach in a nutshell. Within this analytical framework, decision-makers' creative political action depends on their capacity to utilize the scope for what is politically feasible to achieve in the interest of their country as they perceive it. What is politically feasible in foreign policy depends partly on structural opportunities and limitations in the global and domestic environments of the decision-making state, including the alignment of interests with other actors. However, the perception of political feasibility also depends on the quality of decision-making institutions and procedures, the willingness of individual decision-makers' to accept a risk, as well as on their capacity to influence the construction of narratives upon which alliance partners and foes make or justify decisions.

Such framed, FPA very much acknowledges that human agency is not limited to the passive or mechanical adaptation to structural circumstances, even if most tasks of foreign affairs and security are dealt with as matters of routine at an appropriate level of management. On issues where the political stakes are high, we assume that decision-making agents proactively attempt to exploit favorable structural conditions to pursue their political aims, whether the mechanism at work is the single statesmen-like decision, bureaucratic negotiations, or the result of some brilliant standard operating procedure (SOP) fitting the purpose extremely well. The political engineering gene in FPA may be expressed in all sorts of ways.

FPA is a state-centered, bottom-up approach to the study of caveats. Recall that we selected FPA as the "hard core" of the empirical research program precisely because the phenomenon of caveats follows from a foreign policy decision and because FPA is fit to inspire and guide empirical research on complex decision-making and implementing processes. Even if FPA is state-centered, no simplifying assumption is made of the state as a unitary actor. Quite the contrary, FPA tears down the walls in the proverbial "black box" of decision-making imposed by most rational IR approaches to prepare for the empirical investigation of foreign policy-making processes in complex social organizations.

As crucial, FPA invites the study of how the global and domestic environments of the foreign policy-making state both influence decision-making processes and subsequent outcomes regarding implementation. How, at what stage in the decision-making process, and to what extent are empirical questions. FPA is thus also a multi-level analytical approach to the empirical tracing of policy-making and implementing processes in the broad domain of the politics of foreign policy-making. Presumably also capable of approaching decision-making as to whether to apply caveats, how, to what extent, and as a substitute for what? In the political study of coalition participation and caveats, FPA would seem to be particularly well equipped to shed light on research questions such as:

- What decision-making problem may application of caveats address for foreign policy-makers confronted with the prospects of committing troops to a coalition force?
- Why do some states choose to apply caveats on their troop contributions, while other members of the alliance decide to renounce from participating in any military capacity?
- How may a better understanding of the foreign politics of caveats improve our understanding of the dilemmas of alliance politics and the military implementation of political mandates of coalition forces?

As explained, our programmatic study does not pretend to offer definitive answers to such questions. That is for subsequent empirical research to do. However, we should expect the approach of FPA to provide clues as to *how* we may arrive at credible and useful answers to such kinds of research questions, partly by conceptualizing the foreign policymaking process, and partly by inspiring and supervise the borrowing of explanatory theories from several levels and bodies of political analyses.

As indicated, the key tasks in foreign policymaking are related to the assessment and influencing of the scope for political maneuvering (SPM), the context-specific interpretation of the national interest, the choice and calibration of packages of policy instruments, and the development of justification rhetoric capable of reducing the political costs of controversial decisions. In foreign policy and in alliance politics, where stakes are high, political communication is neglected at our peril. As for *research* on policy-making processes, it is crucial to gather data on each one of the decision-making phases and the most promising framework conditions and agents influencing decision-making and outcomes. No small task.

The FPA approach is then an eclectic, theory borrowing, multi-level, and decision-making emphasizing approach. Having reasoned the epistemological function of FPA for the empirical research program in the previous chapter, in the present chapter we discuss the agency of foreign policy decision-making, the concept of foreign policy in relation to caveats as a policy instrument, and the ideal-typical stages in foreign policy-making and implementing processes, while keeping in mind that "the reality of policy-making is extremely messy" (Clarke 1996: 27).

DECISION MAKERS, FOREIGN POLICY, AND FOREIGN POLICY-MAKING

The principal decision-makers within the FPA approach are the strategic foreign policy decision-makers embedded within the decision-making and implementing institutions of the state. For our purposes, decision-making agency rests with foreign policymaking actors responsible for making decisions as to whether and how support coalition operations, the question of national reservations on the use of force included. The decision group will include, at some point, the head of government and a small handful of cabinet members (the ministers of defense and foreign affairs, in particular), as well as key civil servants and advisers from the ministries of foreign affairs and defense.

Military officers are included to provide policy-relevant information and advice on the feasibility, planning, and implementation of operations. If the process concludes with a decision to participate in coalition force, the political decision-makers delegate the responsibility of implementing the policies to the relevant agencies of the state, predominantly the military. As pointed out in Chapter 3 and further theorized in Chapter 10, the arrangement of principal–agent relationship between political decision-makers and operational implementers as to political control over the sword may vary considerably (Miller 2005).

In some coalition forces, governments delegate considerable discretion down the chain of command in deciding how the regulation of the use of force (RoE) be interpreted. This is more likely to happen in instances where local knowledge considered a more crucial source of situational judgment than general/prior knowledge. In other instances, governments make arrangements to secure tight political control over the operational implementation. Arguably, this is more likely to happen in a highly politicized environment, where general knowledge is a prerequisite for making competent decisions on the use of force. As argued in Chapter 4, the extent to which national decisions on the use of force in coalition operations is delegated down the military chain of command is a key empirical question in the study of the politics of caveats (Auerswald and Saideman 2014).

In FPA, we make no fundamental distinction between different policy areas of foreign affairs, such as trade policy, foreign aid policy, immigration policy, human rights policy, and defense and security policies. While the politics of caveats belongs to the latter, the approach of FPA is capable of coping with different empirical peculiarities of foreign policy-making within the same framework. That is, provided that the choice of supplementary theories ("auxiliary hypotheses") and data are relevant for the particular foreign policy phenomenon under scrutiny.

We may define foreign policy and the art of foreign policymaking in different, but complementary ways. At its most general, Philip Alan Reynolds defines foreign policy as "that set of actions which is executed by governmental agencies toward other states on the international arena [...] with the aim of advancing the national interest" (1994: 39). Charles W. Kegley and Eugene R. Wittkopf refer to "foreign policy and the decision-making processes through which foreign policy is created, [as...] the goals public decision-makers seek to achieve outside the borders of the state, the values underpinning the objectives, and the means applied to achieve them" (1997: 40). This definition distinguishes clearly between the "politics" and "policy" dimensions of foreign policy. The first relates to the process of making decisions, and the latter to the tangible results of the policy-making process as expressed in negotiated aims, means, and actions.

For some research purposes, it may be useful to conceive of foreign policy as the state's projection of self-interest on the global arena, within the SPM perceived to exist and constructed through creative political action (Fermann 2013: 13). Foreign policy may be expressed proactively as the domestically generated projection of national interests, or reactively as the externally induced adaptation policies (ibid.: 47). More specifically, we may think of foreign policy in terms of the outwardly directed and purpose-oriented enterprise of the territorial state, where strategies are selected, and instruments are applied in light of the state's collective self-conception (identity), foreign policy goals, capabilities, and the specific challenge that face the state (ibid.: 91).

In stating that "foreign policy is that area of politics which bridges the all-important boundary between the nation-state and its international environment" (Wallace 1971: 7), William Wallace gives a reminder that borders still are significant features in an interdependent/globalized world. This is also the reason why it is justified to speak of foreign policy-making as pinched between a rock and a hard place. That is, between the often conflicting demands of global and domestic politics.

Finally, we have, in tune with the FPA approach, suggested that foreign policy-making is about identifying the scope for creative political action, prioritizing national interests, create instrumental cohesion between goals and means, and the communication of justifying arguments that contribute to the construction of winning political coalitions and reduce the loss of political capital. Indeed, political capital is necessary to preserve and grow for the rainy day when political support is required to defend another difficult decision (Fermann 2013: 83).

As emphasized in Chapter 4, constructing definitions goes beyond the thrills of the analytical exercise. Making analytical decisions on the nature of the phenomenon of foreign policy informs two related, but epistemologically distinct domains: One concerns the factual question as to what are the main tasks and functions of the strategic foreign policy-makers. The other relates to how fine-tuned the operationalization of foreign policy as dependent variable needs to be in an analytical framework such as ours.

The definitions presented above clearly indicate that the key decision-makers are responsible for several interrelated tasks. In an ideal-typical decision-making situation, the ministry of foreign affairs should act as they pretend to do: Check if it is within the capacity of the state to do something, anything. If so, consult what interests can be served within the SPM revealed. Decide what interests to pursue within the limits of capacity. Be clever about the choice and calibration of policy instruments serving the foreign policy preferences in question. Implement the instrumental decisions made. Observe in Fig. 6.1

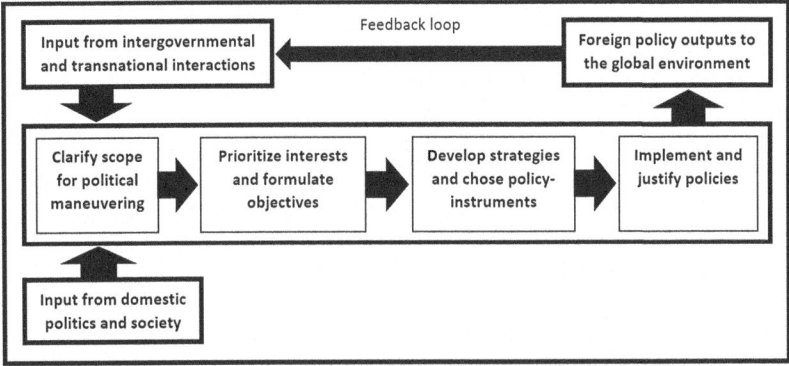

Fig. 6.1 Foreign Policy Analysis—modeling the foreign policy decision-making and implementing process

that implemented foreign policies that are responded to in the global realm feed back into another cycle of the foreign policy-making process. We may imagine a similar feedback loop via domestic reactions to a government's foreign policy.

However, real-life decision and implementing processes may not be as streamlined as depicted in Fig. 6.1. As mentioned, political decision-making is sometimes highly irregular. Policies expressed in goals and instruments may fail to implement, and the seeming lack of coherence between stated goals and policy instruments may be because the real motives for the policy are not publicly stated. Also, the actual implementation of adopted policies (goals and means) may be opposed by some party that lost out in the preceding struggle on policies. Still, it makes sense to model the foreign policy decision and implementing process along the lines of an ideal-type construct.

More than two decades ago, Kaleb Holsti asked, "at what point [in the causal chain] does foreign policy become international politics?" (1995: 18). Analytically speaking, foreign policy starts when inputs from the global and domestic environments hit the papers or the desk of some embassy or ministry. It ends when the output of foreign policy-making, preferences, and policy instruments expressed in foreign policy behavior becomes part of the international exchange between states. At that precise moment, foreign policy morphs into international politics. A new cycle of foreign policy-making starts when the feedback loop from international politics hit the desks of decision-makers as other governments' reaction to previous foreign policy actions or reactions—speech acts and calculated in-actions included (Fermann 2013: 100–102).

CAVEATS IN FOREIGN POLICY-MAKING

Recall that caveats were defined as national reservations on the use of force in coalition forces. In the context of FPA, this definition elevates caveats to the broader context of foreign policy instruments at the disposal of the state, as discussed in Chapter 2. However, foreign policy is not limited to the instrumental aspect of decision-making. Foreign policy also includes the negotiation of what is the SPM, foreign policy preferences and goals, material actions and implementation of policies, and particular speech acts executed to justify foreign policy actions.

Subsequently, we will use this fivefold specification of the foreign policy-making and implementing process as knobs to hang our discussion of caveats statecraft. For now, we emphasize that the study of the chain of foreign policy-making invites a multitude of caveats-related research questions and avenues. We may study caveats in terms of SPM. Does the policy instrument of caveats widens the repertoire of what is politically feasible to do in foreign policy and in alliance politics? Furthermore, what particular foreign policy objectives are served by the application of caveats? What larger package of policy instruments does consideration of caveats enter into? To what extent are non-declared caveats applied? What types of caveats are applied and under what conditions? What kinds of caveats are most destructive to alliance politics and military effectiveness?

Note that some of the research questions are limited to the mapping of a single variable. In other cases, the ambition is explanatory and requires the research of relationships between variables. Picking up on the probability that decision-making processes do not always "behave" and turn out according to the coherence and the sequencing assumed in ideal-typical models of political processes, we may want to explain the discrepancy observed between different decision-making tasks in foreign policy. Why did the government fail to implement the policy instrument (say, caveats) it had decided on to apply? Why did not the policy instrument applied (say, caveats) resonate well with the government's publicly stated aim, say, a strong military footprint in support of the mandate of the coalition force? Why did the government fail to contribute militarily to the coalition force, even if it was plausible that a military contingent with restrictive caveats was inside the government's SPM?

Finally, we would like to emphasize that modeling the foreign policy-making and implementing process, and the input and output environments as different parts of the same causal chain, implies that what is one researcher's dependent variable may become another researcher's independent variable. The epistemological status of any one of the intermediate links in the decision-making and implementing process depends very much on the research question. If the research question limits the study to map or explain foreign policy intent, foreign policy behavior is not even part of the research. However, if we want to explain the actual foreign policy behavior of a state as regards caveats, the latter phenomenon is the dependent variable. The causal chain of decisions on SPM, preferences and policy instruments, are intermediate variables (if undocumented, only theorized mechanisms) for the global and domestic independent variables assumed to condition the particular foreign-policy behavior in the first place. The take-home message that cannot be stressed too much is that analytical constructs should not be accepted as a straightjacket, but adapted with the flexibility to facilitate particular research needs.

We have shown how the phenomenon of caveats may be framed in an ideal-typical model of the foreign policy-making and implementing process, very much in line with spirit of the FPA approach. Furthermore, we have argued that the operationalization of foreign policy needs to be belayed in a definition of foreign policy, which captures the main functions of foreign policy-making. Now, we venture into what it means to take political responsibility for the foreign policy-making process, also as relates to whether and how to contribute to coalition forces. Knowledge of the substantial content of decision-making processes is at the core of FPA.

FOREIGN POLITICS AND TASKS OF STATECRAFT

In an interdependent world where most ministries and governmental agencies nurture relationships with similar bodies in other states as well as with intergovernmental agencies, it is left to the ministry of foreign affairs to create some cohesion in foreign policy. This coordination function comes on top of the strategic foreign policy decision-makers' main tasks related to the assessment of SPM, the prioritization of national interests, the choice of policy instruments, and the justification of foreign policy behavior.

These tasks are critical whether the issue is some major reorientation of security and defense policies (say, from the territorial defense, to "out-of-area" operations), or a particular crisis-situation with much at stake (say, an incident related to dispute over territorial claims). Typically, decisions related to the participation in coalition operations qualify for the careful attention of the strategic foreign policy-making leadership (Fermann 2013: 51).

We need to discuss the challenges involved in making foreign policy decisions because the generic dilemmas encountered and the trade-offs made in foreign policy-making also apply to decisions made on the issue of caveats. Also, any discussion of political decision-making needs to start from a realistic conception of the nature of politics. Politics is the competition for gaining access to and use positions of power, governance, and authority to influence the distribution of scarce goods and inconvenient burdens within and between societies, groups, and institutions. Within well-functioning states, the competition for decision-making positions and the access to decision-making processes is regulated by constitutions and laws that are enforced (Easton 1965: 50).

Politics in this sense is about winning the right to govern the state as a legitimate practitioner of power within a delimited territory and, in states with a democratic form of government, for a limited period (Reynolds 1994: 10–11). Typically, politics is expressed in "elections, legislation, interest group activity, government formation, and decision-making in government, but may also in a wider sense include the exercise of power and authority within private organizations, neighborhoods, small groups, families and relationships" (Østerud 1995: 15).

In international relations, politics is essentially unregulated, although normative institutional frameworks do exist which may moderate the international anarchy. Still, where much is at stake, and interests sharply diverge between states, might usually convert into "right." Indeed, the power footprint of the hegemonic power(s) is already visible in the codification and interpretation of International Law and international regimes that provide the normative basis for any attempt at regulating international politics.

Political man compete for a host of goods and resources, which are scarce, contested and sometimes indivisible in structure:

Security and sovereignty; strategic resources such as water, fertile soils, livestock, minerals, and energy sources which may be converted into other goods; attention, time, prestige, and posthumous reputation; means of payment, privileges and rights; ideological hegemony; values which are to be held in high esteem; competence and knowledge; and life prospects and living conditions. We will come back to how these motivational urges may translate into foreign policy interests and goals below.

On the global arena, states are the predominant political form. Through foreign policy decision-making and implementing processes, states project their interests toward the global arena and thus participate in the international struggle for scarce goods and sought-for positions, material as well as nonmaterial. In preparing for this competition, the first task of the strategic foreign policy-makers is to assess the scope for political action on the issue in question. For instance, whether it is within the resources of the state and politically feasible to contribute to the coalition force under establishment. If so, on what terms?

Caveats and the Scope for Political Maneuvering

The metaphor and concept of "scope (or space) for political maneuvering" (SPM) we reserve for the set of policy options which is perceived to be politically feasible and operationally available for the strategic leadership after external and domestic limitations and opportunities have influenced the foreign policy issue at hand. It is essential that the strategic foreign policy-making leadership sort out how to assess and exploit what is politically feasible to do under prevailing circumstances. Also, the key decision-makers need to consider how SPM be extended by influencing relevant framework conditions and actors.

SPM may be extensive or minimal, but not unlimited. This is precisely because international politics is the competition for goods and positions in a realm inhabited by other agents (states, organized groups) with the capability and will to influence the outcome of interactions for their own benefit. If a state's SPM is close to zero, this implies that the strategic policy-makers for all practical purposes are reduced to administrators of political necessity. In such a situation of political impotence, the decision-makers have not merely lost the political and operative initiative. It also implies that the policy options are very limited, or even absent, within a less ambitious strategy of adaptation.

For instance, after the terrorist attacks of 11 September 2001, most allies of the United States were sympathetic toward The Bush Administration's call for a forceful retaliation—the "War against terror." However, even the more skeptical ally would have found it outside the foreign policy SPM to deny the request for some material or political support. Also in the procurement of expensive weapon platforms, it would probably be outside the foreign policy SPM of a small ally to buy fighter jets from outside the sphere of the alliance (Aaberg 2016).

It is demanding to establish with a degree of certainty whether a foreign policy decision is a decision made out of necessity or one made out of independent choice. In particular, if the decision relates to high-stake security issues where political deniability and operational secrecy may be critical. In Chapter 11, we elaborate on issues related to the gathering of data on security policies and operational implementation. Regardless of the methodological challenges of data collection, foreign policy decisions are assessed on the merits of the SPM available at the time, and not on some unrealistic notion of what is possible to achieve through foreign policy-making.

Everything else being equal, smaller states have a much slimmer margin for survival in international affairs than have more powerful states. This limits the risks the strategic leadership of small states is willing to take. The more crucial it is that smaller members of alliances assess the SPM correctly, both in the more general trade-offs made between fears of entrapment and abandonment in alliance politics (Snyder 1984), and in the decisions made on coalition participation. Indeed, one interesting avenue of research would be to study the notion that smaller allies' are more likely to apply caveats on their contribution to coalition forces.

In shedding light on the SPM, the foreign policy strategic decision-makers need to take several steps: First, the decision-makers gather information on the challenge ahead to make sure their perception of the problem based on relevant facts. This is not only about gathering sufficient and relevant data. Decision-makers also need to mobilize the relevant expertise to interpret data in meaningful ways; for instance, on the implications for security policies, participation in coalition forces, and the application of caveats. For this purpose, the strategic leadership in the ministries of foreign affairs and defense draw upon resources in the security agencies, the military, the police, and even researchers and consultants (Ruud 2010; Arntzen 2010; Tangen 2010; Kynø 2010).

Second, strategic decision-makers are challenged on their capacity to relate knowledge on the decision-making problem (say, an external threat to national security, or a request to participate in a coalition force) to own interest perceptions (say, the balancing of concerns about entrapment and abandonment), response options (capabilities, available policy instruments), and willingness to take risks. It is in this complex decision-making crossroad that the extent and the quality of the SPM become more evident.

Indeed, it is precisely here that the outline of the preferred, possible, too risky, and utterly unrealistic policy options are processed. It is also at this decision-making stage that the catastrophic miscalculation or the stroke of genius is conceived. Scholarship on international political history provides ample of evidence on strategic decisions that have been instrumental in changing the trajectory of history (Roberts 2007). Much more frequent, however, are decision-making situations with less at stake. Regardless, strategic decision-makers are assessed on their ability to make use of the possibilities for taking creative action within the parameters of preferred interests, acceptable risk, and what limitations prevail.

Finally, the competent strategic leadership will look for ways to extend the SPM. This may result from being better informed about the conditions for political action, not least by gathering additional information on the opponents' capabilities and intentions. The perception of SPM can be altered by influencing key conditions for the decision-making situation. If a state manages to build lasting or ad hoc alliances with other states, this may increase the SPM and the repertoire of policy instruments substantially.

Moreover, the strategic leadership may influence (increase) the SPM through how they present the case in question toward the public opinion and parliament. Politicians know very well, as do marketing consultants, that the conception of reality is not limited to singular definite form. Rather, the conception of reality is subject to social construction (Wendt 1999). This implies that decision-makers' conception of reality and what they deem relevant information for the decision-problem at hand, are subject to the active and calculated manipulation for political purposes. In much the same way as a video vignette may influence potential customers attitude toward a product, the narratives and arguments strategic foreign policy decision-makers construct may influence other parties' and opponents perception of the situation and, in turn, their behavior.

The assessment of the SPM will also depend on the eyes that see, and the minds of those who assess the SPM. In particular, the assessment of the SPM will depend upon how strategic decision-makers deal with uncertainty and risk. Invariably, complex political decisions are made without full information, or without the capacity to fully process available information. Uncertainty in the assessment of goal–means relationships reduces the probability that a policy instrument will serve the purpose (policy goal). This, in turn, increases the probability that the implementation of policy instruments lead to unanticipated and undesirable outcomes. This is the core problem of making decisions under uncertainty.

The willingness to accept risk is probably associated with the perception that the SPM is increasing. Everything else being equal, risk-prone policy-makers are likely to perceive that they have more freedom of action than risk aversive policy-makers. Risk is related to uncertainty but is not identical. A risk is a function of the probability for a bad outcome and the harmfulness of the consequences of this outcome.

For small states, in particular, there are reasons to be cautious about how much risk to accept, if the political risk is possible to quantify in a meaningful sense at all. However, when the state's core values and survival are at stake, also small states must take great risk to protect its key interests and survival (Levy 1996, 1997). The implementation of extraordinary policies may presuppose the willingness to accept risks and substantial losses in the case policies should fail.

On the question of risk willingness, there are two extreme answers: One is untamed arrogance and hubris. The other is to become a hostage of one's fears. An extremely risk-prone decision-maker is in danger of making reckless decisions that will cost dearly. An extremely risk aversive foreign policy decision-maker will gravitate toward political servility and the de facto dismantling of political independence. On the issue of risk-taking, the strategy of the happy mean is therefore likely to be the better option.

The question remains, however, what the happy mean is to imply when a lukewarm coalition partner are invited to contribute to a coalition force. The policy instrument of restrictive caveats may become a crucial part of the answer. Another educated guess would be to propose that a risk-prone leadership group is more likely to contribute troops to a coalition force without any restrictive caveats attached. Finally, we might expect that a government less concerned about abandonment, and seriously concerned about entrapment and entanglement in out-of-area operations to limit their contribution to political and economic support. We elaborate on Jack S. Levy's prospects theory on the willingness to accept risk in Chapter 10 to explain caveats behavior.

Caveats and Foreign-Policy Preferences and Instruments

Having assessed and maximized SPM, the strategic foreign policy decision-makers are in a position to decide for what purpose and how freedom of action be exploited. That is to say, what foreign policy goals served using what foreign policy instruments? In the previous section, we argued that the perception of SPM is likely to influence decisions on coalition participation and caveats. In Chapter 2, the phenomenon of caveats was classified in the context of the wider portfolio of foreign policy instruments. Recall that caveats belong to the category of foreign policy instruments related to the threat and the use of lethal force. Caveats either subtract or add to the foreign policy instrument of lethal force, depending on whether the caveats are restrictive or permissive.

Caveats may relate to other instruments of foreign policy as well. First, the extent of caveats-use likely is communicated mainly through diplomatic channels, and in multilateral negotiations on burden sharing. Second, just as we have seen instances of alliance partners paying themselves out of military commitments altogether, side-payments of different sorts be offered to compensate the alliance for the inconveniences resulting from the use of restrictive caveats in coalition forces. Moreover, a caveats-applying government (restrictive caveats) may show to military investments in other coalition operations to justify a less than full commitment in the current coalition force. In the first case, caveats are related to diplomacy as a foreign policy instrument. In the second instance, caveats are related to economic compensation. In the final case, political rhetoric is applied to explain caveats-policy in terms of the burden-sharing record.

As to the question of what foreign policy interests and goals caveats serve, the simple and general answer is "some national security interests." National security relates to the survival of the state and is the foundation of any national needs-pyramid on which economical, ideational, and other interests rest. In Chapter 4, we argued that the concept of caveats be reserved for restrictive and permissive national reservations on the use of force in the context of coalition forces. This implies that caveats as a foreign policy instrument serve government's security concerns within the framework of alliance politics.

In alliances, small and medium powers pay for increased territorial security with reciprocity clauses, military cooperation, and political loyalty to the leader of the alliance. After the Cold War, Western alliance commitments were extended to solidary participation in out-of-area coalition operations. However, members of the NATO alliance participate in such missions with varying degrees of enthusiasm. This is because out-of-area operations are not about the defense of coalition members' borders and territories in any immediate sense.

Participation in out-of-area operations is supporting the alliance leader in shouldering the responsibility to defend broader and long-term geopolitical security interests. The mission may relate to the defense of international sea-lanes, the pushback of regional challenges to the status quo and international stability, and interventions motivated by a mix of concerns related to regime change, strategic resources, and humanitarian issues. Obviously, participation in coalition forces in Africa, Central Asia, and The Middle East are harder to explain to domestic audiences than defense spending and risk-taking closer to home.

In Chapter 8, we discuss in some detail how the foreign policy instrument of caveats is likely to serve the better balancing of the individual security dilemma in alliance politics that been pushed to the extreme in the distant, costly and risk-prone out-of-area operations in the post-Cold War era, in particular. Glenn H. Snyder (1984) explains this alliance dilemma as the trade-off each and every member of the alliance need to make between (i) its fear of being abandoned from the alliance if not contributing sufficiently, and (ii) the fear of being entrapped in the entanglements of a war in which participation may be motivated more out the need to nurse the relationship with the alliance leader than any perceptions of an immediate threat to national security.

The members of the alliance must weigh their interests and opportunities regarding whether and how to participate. A failure to contribute will have to be explained in the alliance burden-sharing negotiations, at a political cost. The decision to participate with restrictive caveats may strike a better balance between the hesitant ally's fear of being singled out for criticism, and its fear of being drawn into a theatre of war with no clear and immediate individual security interests at stake. On the other hand, we might expect that a government refrains from using restrictive caveats if it (i) considers the threat motivating the establishment of the coalition force to be a serious threat to its national security, (ii) is eager to prove itself in the eyes of the alliance leader, and (iii) has strong reason to be concerned about its political or military reputation within the broader membership of the alliance.

Other broad categories of interests in foreign policy are economic objectives, international prestige and reputation, and different kinds of ideological, religious, and cultural objectives. These interests may relate to the application of the particular policy instrument of caveats in indirect ways, but not elaborated on here.

Caveats and Justification of Foreign Policies

What requires some elaboration, however, is the final task of justifying policies on coalition participation and the use of caveats. The way policy-makers reason, justify and communicate their policies very much depends on whom they want to influence and for what purpose. The application of restrictive caveats is not something a government would brag about in the context of burden-sharing negotiations within the alliance/coalition. Quite the contrary, toward such a target group of peers a caveats-applying government would choose settings and language that downplay the significance of the caveats to minimize the political costs resulting from its reserved contribution to the common effort.

However, in a domestic political context where skepticism toward any military participation might be strong, the government's application of restrictive caveats on the contribution may instead be considered a political asset, yet another instrument to use to extend the scope of what is politically feasible to do. Indeed, restrictive caveats may be precisely what is required to convince a reluctant coalition partner in government, and a hung parliament to side with the government and allow for the participation in the coalition forces. In turn, by referring to this sensitive domestic political situation, the government can fend off critique in burden-sharing negotiations with allies.

Justifying the use of lethal force in foreign policy toward domestic constituencies is among the most demanding task of foreign policy-making. In a handful of studies on Norwegian governments' justification rhetoric in support of military contributions to five Western coalition forces, we found that all sorts of justifying arguments related to UN authorization, operational aptitude, alliance commitments, national interests, and political values were applied in all cases. Although to a varying extent across cases, depending on particular circumstances (Singsaas 2016; Radpey 2014; Johnsen 2014; Hermansson and Fermann 2013; Hermansson 2010). One plausible interpretation of this empirical pattern is that a multitude of justifying arguments are deemed to be required to attract the political support from diverse public audiences and political constituencies in Norway. To stake everything on a single argument would likely not be sufficient to mobilize a winning political domestic coalition for the war effort. For the NATO-friendly political bases of the Conservative and the Progressive parties of Norway, and a large part of the Labour Party, the alliance commitment argument is usually sufficient to mobilize support. As for the Christian People's Party and the Socialist Left, humanitarian justification arguments have much greater political appeal. Across the political spectrum, unambiguous UN Security Council Resolution in favor of military intervention has a reassuring and mobilizing effect in favor of Norwegian participation in the coalition force.

For reasons not fully understood, the question of caveats has not been part of the Norwegian public debate to the prominent extent of the justifying arguments mentioned above. The related debates as to whether and how Norway should contribute to coalition forces have yet to be linked in the public opinion. However, there are indications that the issue of caveats been used as a negotiation card and a political concession to persuade the Socialist Left, a junior partner in two Centre-Left governments, to accept Norwegian contributions to Western coalition forces (Frost-Nielsen 2011).

The lessons learned from research on Norwegian coalition participation are, first, that justification debates on whether to contribute to coalition forces are part of the public domain. Second, discussions on how Norway should contribute, and in particular whether restrictive caveats be applied are limited to a much smaller circle of stakeholders. An interesting line of research would be to investigate what relationship there might be between the pattern of justification arguments used in favor of coalition participation, and the propensity to apply caveats on the military contribution. If any, and why?

In the preceding discussion, we have moved forth and back between both the generic and the singular level, and between the perspectives of the policy-formulating executive and the probing researcher. This double dualism is to some degree reflected in Table 6.1. For our programmatic

Table 6.1 Foreign politics of caveats—policymakers' decision-making checklist, and researchers' data-gathering focus of attention

Key tasks of statecraft	Caveats relevant foreign-policy questions
1. Clarify scope for political maneuvering (SPM) in foreign policy	Does the SPM as perceived allow for, or invite the use of national reservations on the use of force? May SPM be enhanced in ways that change the relevance of national reservations on the use of force as policy instrument?
2. Prioritize interests and formulate foreign policy objectives	What interests, objectives, and goals are served by the use of national reservations on the use of force?
3. Develop strategies and choose policy instruments	What kind of national reservations on the use of force is the most effective, on their own or in conjunction with other policy instruments? If at all feasible.
4. Implement foreign policies	How are national reservations on the use of force to be effectively implemented and administered in the area of deployment?
5. Justify foreign policies	*Toward domestic constituencies*: How to argue national reservations on the use of force to secure domestic political support for participation in coalition force? *Toward alliance partners*: How to communicate the application of national reservations on the use of force in ways that deflect criticism and minimize loss of political capital?

purpose, the probing questions are used as a point of departure for theorizing and research on caveats-related decision-making on any or all of the five steps in the ideal-typical model of foreign policy decision-making and implementation.

REFERENCES

Aaberg, M. (2016). *Kampflykjøp mellom barken og veden. En utenrikspolitisk analyse av beslutningen om å velge F-35 som Norges neste kampflyplattform*. Master Thesis in Political Science, Trondheim, Department of Sociology and Political Science, Norwegian University of Science and Technology (NTNU). https://brage.bibsys.no/xmlui/bitstream/handle/11250/2419669/Aaberg%2C%20Magnar.pdf?sequence=1.

Arntzen, T. (2010). Etterretning og staten. Forholdet mellom produsent og bruker. In G. L. Dyndal (Ed.), *Strategisk ledelse i krise og krig* (pp. 131–140). Bergen: Fagbokforlaget.

Auerswald, D. P., & Saideman, S. M. (2014). *NATO in Afghanistan: Fighting Together, Fighting Alone*. Princeton, NJ: Princeton University Press.

Carlsnaes, W. (2002). Foreign Policy. In W. Carlsnaes, T. Risse, & B. A. Simmons (Eds.), *Handbook of International Relations* (pp. 331–349). London: Sage.

Carlsnaes, W. (2004). Comparative Foreign Policy Analysis in a Historical and Contemporary Perspective. In M. Hermann & B. Sundelius (Eds.), *Comparative Foreign Policy Analysis. Theories and Methods* (pp. 36–63). Englewood Cliffs, NJ: Prentice-Hall.

Carlsnaes, W. (2008). Actors, Structures, and Foreign Policy Analysis. In S. Smith, A. Hadfield, & T. Dunne (Eds.), *Foreign Policy: Theories, Actors, Cases* (pp. 85–100). Oxford: Oxford University Press.

Clarke, M. (1996). Foreign Policy Analysis: A Theoretical Guide. In S. Stavridis & C. Hill (Eds.), *Domestic Sources of Foreign Policies: Western European Reactions to the Falkland Conflict* (pp. 19–39). Oxford: Berg.

Clarke, M., & Brian, W. (1989). *Understanding Foreign Policy: The Foreign Policy Systems Approach*. Aldershot: Edward Elgar.

Easton, D. (1965). *A Framework for Political Analysis*. Englewood Cliffs, NJ: Prentice-Hall.

Fermann, G. (Ed.). (2013). *Utenrikspolitikk og norsk krisehåndtering*. Oslo: Cappelen Damm Akademika.

Foreign Policy Analysis. (2013). *Journal*. Oxford: Oxford Academic. http://onlinelibrary.wiley.com/journal/10.1111/(ISSN)1743-8594/homepage/ProductInformation.html.

Frost-Nielsen, P. M. (2011). Politisk kontroll av militær deltakelse i internasjonale operasjoner. Restriksjoner på bruk av norske kampfly i Afghanistan. *Internasjonal politikk, 69*(3), 359–385.

Hermansson, H. (2010). *Studie av norsk legitimeringsargumentasjon for deltakelse I NATO "Out-of-Area" operasjoner*. Master thesis in Political Science, Trondheim, Department of Political Science, Norwegian University of Science and Technology (NTNU).

Hermansson, H., & Fermann, G. (2013). Myndighetenes legitimering av norsk deltakelse I NATO-operasjoner I Bosnia, Kosovo og Afghanistan. In G. Fermann (Ed.), *Utenrikspolitikk og norsk krisehåndtering* (pp. 335–354). Oslo: Cappelen Akademisk Forlag.

Hill, C. (2003). *The Changing Politics of Foreign Policy*. Houndmills, Basingstoke: Palgrave Macmillan.

Holsti, K. J. (1995). *International Politics: A Framework for Analysis*. Englewood Cliffs: Prentice Hall.

Hudson, V. M. (2007). *Foreign Policy Analysis: Classical and Contemporary Theory*. Boulder, CO: Rowman and Littlefield.

Johnsen, C. (2014). *Kunsten å overbevise. Studie av norske myndigheters legitimeringsargumentasjon for militær deltakelse I Libya 2011*. Master thesis in Political Science, Trondheim, Department of Political Science, Norwegian University of Science and Technology (NTNU).

Kegley, C. W., & Wittkopf, E. R. (1997). *World Politics: Trend and Transformation*. New York: St. Martin's Press.

Kubálková, V. (2001). Foreign Policy, International Politics, and Constructivism. In V. Kubálková (Ed.), *Foreign Policy in a Constructed World* (pp. 15–37). New York: M.E. Sharpe.

Kynø, S.-Fr. (2010). Strategisk etterretningsstøtte, suksesskriterier og forbedringsmuligheter. In G. L. Dyndal (Ed.), *Strategisk ledelse i krise og krig* (pp. 149–168). Bergen: Fagbokforlaget.

Levy, J. S. (1996). Loss Aversion, Framing, and Bargaining: The Implications of Prospects Theory for International Conflict. *International Political Science Review, 17*(2), 179–195.

Levy, J. S. (1997). Prospect Theory, Rational Choice, and International Relations. *International Studies Quarterly, 41*(1), 87–112.

Light, M. (1994). Foreign Policy Analysis. In A. J. R. Groom & M. Light (Eds.), *Contemporary International Relations: A Guide to Theory* (pp. 259–281). London: Pinter Publishers.

Miller, G. J. (2005). The Political Evolution of Principal-Agent Models. *Annual Review of Political Science, 8*(1), 203–225.

Neack, L. (2013). *The New Foreign Policy: Complex Interactions, Competing Interests*. Lanham, MD: Rowman & Littlefield.

Østerud, Ø. (1995). *Statsvitenskap*. Oslo: Universitetsforlaget.

Radpey, A. (2014). *Rettferdiggjøring av maktbruk. Kartleggingsstudie av norske myndigheters legitimeringsargumentasjon for deltakelse i luftkrig i Libya 2011*. Master thesis in Political Science, Trondheim, Department of Political Science, Norwegian University of Science and Technology (NTNU). http://docplayer.me/47634637-Azita-radpey-rettferdiggjoring-av-maktbruk-master-oppgave.html.

Reynolds, P. A. (1994). *An Introduction to International Relations*. New York, NY: Longman.

Roberts, J. M. (2007). *The New Penguin History of The World*. London: Penguin Books.

Ruud, M. (2010). Simulering for systemforståelse og beslutningstrening. In G. L. Dyndal (Ed.), *Strategisk ledelse i krise og krig* (pp. 349–360). Bergen: Fagbokforlaget.

Singsaas, A. (2016). *Argumentets kraft. Et klassifiserings- og kartleggingsprosjekt av norske myndigheters legitimeringsargumentasjon for norsk militær deltakelse i Irak 2014–2016.* Master thesis in Political Science, Trondheim, Department of Political Science, Norwegian University of Science and Technology (NTNU).

Snyder, G. H. (1984). The Security Dilemma in Alliance Politics. *World Politics, 36*(4), 461–495.

Tangen, A. (2010). Politisk sikkerhetstjeneste. In G. L. Dyndal, (Ed.), *Strategisk ledelse i krise og krig* (pp. 149–168). Bergen: Fagbokforlaget.

Wallace, W. (1971). *Foreign Policy and the Political Process.* London: Macmillan.

Webber, M., & Smith, M. (2002). Frameworks. In M. Webber & M. Smith (Eds.), *Foreign Policy in a Transformed World* (pp. 7–104). Englewood Cliffs, NJ: Prentice-Hall.

Wendt, A. (1999). *Social Theory of Internationational Politics.* Cambridge, UK: Cambridge University Press.

White, B. (2004). Foreign Policy Analysis and the New Europe. In W. Carlsnaes, H. Sjursen, & B. White (Eds.), *Contemporary European Foreign Policy* (pp. 11–31). London: Sage.

CHAPTER 7

The Essence of Foreign Policy Analysis (II): Exploiting Political Theory at Multiple Levels of Analyses to Explain Foreign Policy-Making Processes and Outcomes

In the previous chapter, we discussed how FPA conceptualizes political decision-making processes often neglected in comparative analyses of covariation related to caveats, and in research emphasizing burden sharing in alliance politics. By directing attention to states' concerns about (i) identifying, constructing and widening the scope for political maneuvering, (ii) prioritizing objectives and goals, (iii) composing packages of policy instruments to serve preferences, and (iv) implementation of policies, FPA invites us to study actual policy-making behavior. In this way, we are in a position to describe the decision-making processes leading to decisions on coalition participation and the application of caveats.

However, if FPA is to become more than a descriptive and systematizing framework for analysis, we need to infuse middle-range theory into the approach that is capable of *explaining* what is going on in policy-making and—implementing processes, and what external configurations of forces influence such processes and impact subsequent caveats-related outcomes. Indeed, the approach of FPA is extremely ambitious in urging the integration of theory from several levels of analyses. In the explanation of decision-making outputs and implementing outcomes, FPA directs us to include the theoretical input that explains caveats from the levels of global politics, domestic politics, as well as from the institutional and individual level of decision-making and implementation. The FPA approach invites us to investigate how domestic and global politics interplay in influencing the foreign policy-making processes that produce decisions on the participation in coalition forces.

© The Author(s) 2019
G. Fermann, *Coping with Caveats in Coalition Warfare*,
https://doi.org/10.1007/978-3-319-92519-6_7

We are encouraged to research how attributes of the state governmental apparatus and key decision-makers influence perceptions of scope for political maneuvering (SPM), preferences and the choice of policy instruments, caveats included. In an integrated FPA approach, all these levels of analysis are theorized for the promise of discovering how global and domestic causal impulses interact through decision-making and implementing processes to produce policies and outcomes (Fermann 2013: 117–128). This is not to say that we are to include all these levels of analyses and phases of decision-making and implementation in every research project. Imagine an extended matrix juxtaposing decision-making phases with multiple bodies of theory at different levels of analyses! Realistically, the holistic ambition of the FPA approach is a *collective* invitation to push the epistemological envelope.

The FPA approach provides directional advice on how to select and translate bodies of theory from the Political Science branches of International Relations, Comparative Politics, Political Behavior, Public Policy and Administration, and beyond for studying foreign policy decision-making and the politics of caveats. Critical selection is necessary due to the vast range of bodies of literature available in Political Science, Political Sociology, Political Psychology, organizational theory, and so forth. Theory translation is often necessary because the dependent variables of other branches of Political Science are related to adjacent or completely different political phenomena such as international cooperation and conflict, elections and public opinion, and areas of public policies other than foreign policy. Hence, the FPA approach does not seem to pay even lip service to concerns about research parsimony but is due to its ontological realism nevertheless likely to contribute to the filling of some crevasses in the present knowledge on the politics of caveats.

In FPA we thus have to do with decision-making, multi-level, theory-borrowing, and theory-translating approach to the study of foreign policy-making on issues related to coalition participation. This makes FPA an eclectic and theory-pragmatic approach that provides us with considerable latitude to adapt the choice and translation of theories to the phenomenon under study and the particular research question. In the present chapter, we review relevant literature from several branches of Political Science, that may be usefully translated for explaining of foreign policy processes and outcomes. The discussion divided into four sections, representing different levels of analyses—the level of governmental politics, political decision-makers, domestic politics and society, and global politics.

THE LEVEL OF GOVERNMENTAL POLITICS: INSTITUTIONAL AND ORGANIZATIONAL APPROACHES TO FOREIGN POLICY-MAKING

In analyses of the state governmental apparatus as a bureaucratic organization, FPA literature draws upon knowledge of organizational theory from, among others, Max Weber (1920) and Herbert Simon (1957). At this level of analysis, the field of FPA seeks to move into the "black box" that the "rational actor" model (RAM) of policy-making built around what are, in fact, much more complex decision-making processes. Rufus Miles' Law—"*where you stand, depends on where you sit*" (Stillman 1999)—is a truism in the study of public policy and administration, and very much applicable to governmental politics.

Miles' law focuses on the characteristics of the state governmental apparatus as a decision-making organization. As an empirical proposition, Miles' Law is capable of illustrating the deductive movement from theoretical assumptions to the empirical application, also in the context of FPA. The core of Miles' Law is the understanding that decision-makers take stands on cases in ways that are determined by the mandates, collective self-understanding and interests that prevail in the institutions they represent. We should not be surprised if the ministries foreign affairs and defense come up with different views on the question of attaching caveats to a military contribution.

This general organizational theory approach is applied for foreign policy-making purposes in Graham T. Allison's "bureaucratic politics" model (BPM). Here, foreign policy decision-making processes are read as a tug-of-war among institutional actors of varying competence and influence levels in the strategic leadership apparatus (1971: 144–184). These actors may, as indicated in the previous chapter, be officials from the ministries of foreign affairs, defense, security, and trade, but can also come from parliamentary standing committees on foreign affairs and defense, and committees established to deal with crises. BPM is thus an inter-institutional model that shows how a series of partially different mindsets be condensed into a single mindset, with the result that a state may appear undivided to the outside world. In other respects, the BPM is the very antithesis of the RAM (Clarke 1996: 33). This is because it depicts a process where several and in part differing perceptions of reality, interests and roles come into contact. Of course, this multi-headed model of foreign policy-making may produce unsustainable or conflicting outputs. It remains an empirical question.

In the intra-institutional "organizational process" model (OPM), Allison shows how foreign policies can also be affected by the establishment and existence of standard operating procedures (SOP); i.e., programmed procedures for dealing with a specific and externally induced problem within an organization (1971: 67–100). Such SOPs may be applied in relevant ministries, but not least in subordinate agencies charged with carrying out decisions, as operationalized in the rules of engagement (RoE) of military organizations discussed in Chapter 3. The main message of the OPM is expressed in a rephrasing of Miles' Law: "*Where you stand, depends on what contingency you have planned for.*"

Other classical contributions at this level of analysis include Huntington (1960), Schilling et al. (1962), Hilsman (1967), Neustadt (1970), and Halperin (1974). A more recent supplement to the literature is Iver B. Neumann's anthropological study on foreign policy-making (2012) informed by participatory observation of decision-making processes in ministries of foreign affairs.

The Level of the Policy Maker: Psychological and Socio-Psychological Approaches to Foreign Policy-Making

Foreign policy assessments and decisions are made primarily on the group and individual level. Given that the approach of FPA was developed in part as a reaction against the simplifying assumptions in the field of IP regarding the state as a unitary and rational actor, it is not surprising that the field of FPA produces much research on the level of actual decision-making to trace policy-making processes (politics) and the outcomes of decision-making processes (policies, implementation). Key contributions on how group dynamics influence foreign policy decisions have been offered by Snyder et al. (1954), Paige (1968), Janis (1972), Hermann and Hermann (1989), Breslauer and Tetlock (1991), Boynton (1991), Khong (1992), and Hart et al. (1997).

When the focus shifts from the state governmental apparatus as an organization via decision-making groups, to decision-makers with certain emotional and cognitive attributes, the study of foreign policy-making borrows insights from Political Psychology. Valerie M. Hudson soberly states that the brain of a foreign policy decision-maker is not a blank slate (2008: 20). Decision-makers bring ideas, values, and self-conceptions that been internalized from their own lives and previous experience, and which form their perceptions of threats, scopes of action, national interests, and preferred paths of action (Sprout and Sprout 1956, 1957, 1965; Levy 2003). Our previous discussion on the decision-makers willingness to take risks largely belongs at this individual level of analysis.

The insight that individual attributes might impact decision-making, even in highly formalized organizations, triggered significant research into understanding how psychological mechanisms affect foreign policy-making processes. How do stress, uncertainty and social standing (De Rivera 1968), "operational codes" (George 1969; Johnson 1977), "orientations" (Hermann 1978; East et al. 1978), and overwhelming or inadequate information (Steinbruner 1974) affect foreign policy-making processes and outcomes? Such questions have been explored comparatively (Singer and Hudson 1992), and within a social-constructivist epistemology (Onuf 1989; Adler 1997, 2002).

The Cold War and the threat of nuclear war between the United States and the Soviet Union was the context for much research on cognitive sources of misconception, and assessments of the intentions of heads of state in rival polities. Key contributors to this literature are Jervis (1976), Cottam (1977), Larson (1985), Lebow and Stein (1990), Walt (1992), and Herrmann (1985, 1993). Research on "limited rationality" and "satisfying" solutions (Simon 1947, 1985), "cognitive biases," (Heuer 1999/1978–86), "heuristic errors" (Kahnman et al. 1982), "muddling through" (Lindblom 1959), and cognitive maps and forms (Shapiro and Bonham 1973) are other important contributions on the level of the individual policy-maker.

On the individual level of analysis, we may reformulate Miles' law as follows: "*Where you stand, depends on who you are.*" Or like this: "*Where you stand, depends on how your mind works.*" Regardless of how little leeway the foreign policy-making institutions offer for variations in individual qualities, the ideas, values, inclinations, cognitive limitations and preferences of policy-makers will—to varying degrees—influence how policy-makers perceive and act upon information. We may use such insights to analyze how "red flag-holders" apply situational judgement in administering restrictive caveats in the field, and to research how policy-makers interpret the SPM concerning participation in coalition forces.

THE LEVEL OF DOMESTIC POLITICS AND SOCIETY: POLITICAL BEHAVIOR, COMPARATIVE POLITICS, AND FOREIGN POLICY-MAKING

Former Speaker of the United States House of Representatives (1977–1987) Tip O'Neill formulated the thesis that "all politics is local" (O'Neill and Novak 1987). This is the notion that political decisions reflect the support that can be gained from society and the pressures executed by political interest groups. In foreign policy, recall that the SPM to pursue domestic interests and impulses is never unlimited, due to prevailing structures in the global environment and the competition between states for all sorts of resources and gains. At the same time, it would be unreasonable to expect foreign policy-making processes to invariably be reduced to clever adaptations to and administration of political necessities imposed from the global political environment.

Ole-Martin Dale (2005), for instance, concluded that domestic political issues were critical factors in the Norwegian governments' decision to decline the US government's request to participate in the invasion of Iraq in 2003. Although declining an opportunity to demonstrate alliance solidarity oftentimes is neither simple or desirable, a small-state government can nevertheless set terms for the use of contingent. Per Marius Frost-Nielsen (2009) shows that even small states have some SPM when large alliance partners request military support in an area of conflict (Afghanistan).

To a certain extent then, Dale and Frost-Nielsen's empirical findings confirm the thesis of "der Primat der Innenpolitik." This is the application to foreign policy of Tip O'Neill's general dictum that the social foundation of politics (society, organized interests) is important if we are to understand political decision-making processes and outcomes. It was German historian Eckart Kehr (1973 [1927]) who coined his main conclusion in the thesis of "der Primat der Innenpolitik" having found that important characteristics of the German naval armament before First World War be traced back to the social structure and organization of financial interests in Germany rather than to an assessment of the external threats facing Germany at the turn of the twentieth century.

The comparative approach to FPA has focused on foreign policy events according to the formula "who does what to whom and how?" (McGowan and Shapiro 1973; Hudson 2008). Typically, the comparative study of foreign policy behavior pays attention to the use of foreign policy instruments

(Gerner et al. 1994; Schrodt 1995; Stavridis and Hill 1996). Research primarily seeks domestic inside-out explanations of variations in foreign policy output and outcome that are observed in the respective states' constitutional structure and form of government (Russett and Shye 1993; Herrmann and Kegley 1995), level of political fragmentation and elite consensus (Hagan 1995), size and level of development (East et al. 1978; Morse 1976; Rummel 1979), financial system (Richardson and Kegley 1980; Katzenstein 1985), and public opinion (Almond 1950; Campbell et al. 1964; Rosenau 1967; Beal and Hinckley 1984).

Case-oriented and conceptual studies have been conducted on the relationship between culture and foreign policy behavior through several of the psychological and socio-psychological mechanisms discussed at the individual level of analysis (Almond and Verba 1963; Pye and Verba 1965), and on the relationship between national cultures of security and foreign policy behavior (Johnston 1995). Constructivist studies of relevance to the field of foreign policy studies have not least looked at how national identity shapes the state's way of viewing its international environment, and thus influencing foreign policy behavior (Walker 1993; Wendt 1994, 1999; Ringmar 1996; Hopf 2002).

The inside-out approach to foreign policy-making is based on the understanding that national experiences, societal institutions, financial matters, and collective self-understandings, directly and indirectly, exercise influence on states' foreign policy. Research on Political Behavior identifies four channels through which social and economic forces exercise influence on public policies in general, but also foreign policy: through political elections (numerical channel); interest group politics (lobbying, corporate channel); public opinion (the media and social media channel); and issue-specific actions and demonstrations (action channel) (Dahl 1973; Hill 2003: 219–249).

Within the multiple channels of political participation, several studies are of relevance to foreign policy-making (Moravcsik 1993, 1997; Putnam 1988; Evans et al. 1993; Skidmore and Hudson 1993; Kaarbo 2015). Acknowledging that the decision-making state is a representative institution, Andrew Moravscik adopts insights from research on Political Behavior. He argues that powerful interest groups influence states' foreign policy preferences and strategic calculations through different mechanisms (1997: 513). On this backdrop, we may rephrase Miles' Law as follows: "*Where you stand, depends on the preferences of your domestic constituency.*" This version of Miles' Law resonates, for instances, with the proposition that restrictive caveats are more likely to be applied on military contributions to coalition forces if key political constituencies are skeptical toward the mission. In particular, restrictive caveats might result if the government is squeezed between a critical opposition at home and an insisting alliance leader demanding a substantial military contribution. We may further speculate that if the critical opposition at home is not balanced by a strong external pressure to contribute militarily, the government would be more inclined to participate in the coalition force in a symbolic capacity, or not at all.

The states' strategic decision-makers cannot create a credible foreign policy in the eyes of the global political environment with open and vocal disagreements at home. Unified foreign policy is a necessary condition for credible foreign policy. This is the main reason why foreign policy-makers go at lengths to build and maintain winning political coalitions around foreign policy objectives and instruments, based on zones of ideological overlap, shared perceptions of threats, as well as the more pragmatic assessment of what is appropriate at the time. In his model of international negotiations as a "two-level game," Robert D. Putnam has dealt with this foreign policy challenge (1988): He starts with the reasonable premise that a necessary condition for entry into international agreements is that sovereign states see that they benefit more from the agreement that is presented than with no agreement at all. In this, Putnam agrees with Andrew Moravscsik (1997) in pointing out that entry into an agreement between states presupposes the establishment of prior agreement within each state.

A domestic SPM must be developed, or "win-set" in Putnam's conceptualization (1988). The size of the win-set determines the flexibility afforded the government by domestic politics at the international negotiating table—be it a matter of contributing military forces to NATO operations, negotiating amendments to the international climate change regime (Fermann 1997), setting up the post-Second World War international regimes on oil-trade and civil aviation (Milner 1993), or to coordination with the EU. At this point, we can see the interaction between domestic and global political variables, and thus between different levels of analyses.

The study of Political Communication may shed light on how domestic political interest are transformed into policies and subsequently defended using political rhetoric (Nacos et al. 2000; Selb 1996). As mentioned in the previous chapter, justification rhetoric is one of several policy instruments available for the building of winning political coalitions (win-set, political consensus) for participation in potentially controversial and risky military coalition operations. It is also a way of spreading political risk: the broader the political foundation for decisions, the lower the political cost to the government in defending the engagement when civilian and military casualties reported in media.

For both purposes, it is a matter of constructing justification arguments that—taken together—may reach, convince, and mobilize active or passive support from a range of fairly diverse, but critical veto groups (Vaara and Tienary 2001; Goffman 1966). The best political result one can aspire to in such cases is a near consensus regarding the policy instrument (say, caveats), based on fairly heterogeneous premises as to goals. As different target groups are receptive on different rhetorical "frequencies," and are convinced by a range of different justification arguments, multi-spectrum justification rhetoric is required to cover arguments based on realpolitik, idealism, and legal and operative appropriateness (Hermansson and Fermann 2013; Fermann 1994).

Such empirical findings are very much in agreement with the following paraphrasing of Miles' Law: "*Where you stand, depends on whom you need to convince.*" Different arguments are required to convince groups with varying interests, affiliations, values, and perceptions of reality. In line with this approach to political communication, foreign policy decision-makers will adjust the justification rhetoric to promote support among the different target groups they need to convince and win over. For instance, on the matters of whether and how to participate in coalition operations.

The Level of Global Politics: International Relations and Foreign Policy-Making

We have repeatedly pointed out that the state as a foreign policy tool of governance must deal with two environments—the internal sentiments and organized interests in the society in which state is embedded, and the external, global political environment. Furthermore, we have insisted that the study of foreign policy is not limited by sterile assumptions about the state that prevails within rationalistic IR approaches. Still, structural characteristics of the foreign policy state's global political environment are crucial for the understanding of foreign policy—more so than ever in the age of globalization. How can insights from the study of IR and International Political Economy (IPE) that global political, regulatory, and economic structures are limiting and enabling framework conditions on the state's foreign policy scopes of action, be incorporated into FPA?

As discussed, structural theory on global politics may be adopted for FPA purposes primarily by abandoning the assumption that the state is a unitary actor. Furthermore, we should consider the state's foreign policy objectives an empirical question, not merely make assumptions. Once translated for FPA purposes, a number of perspectives from IR on intergovernmental anarchy and power polarity (Waltz 1979; Mearsheimer 2001), interdependence (Keohane and Nye 1977), international regimes (Keohane 1989), and International Society (Bull 1977; Bull and Watson 1984) be used to analyse how global political and economic structures impact foreign policy-making processes and outcomes.

For instance, Norway participated in at least seven military operations abroad since the end of the Cold War: in Bosnia, Kosovo, Afghanistan, Libya, and Syria. How can this mobilization of the state's most lethal instruments of power outside the homeland be explained based on a real political outside-in perspective? In several of these missions, Norway has applied restrictive caveats on troop contribution. How to explain this behavior from an outside-in angle? An obvious start is to look at the international threat situation: what was it about the conflicts in Bosnia, Kosovo, Afghanistan, and Libya that may have posed a threat to Norwegian interests and values so seriously that it would motivate the use of armed force? Why did not Norwegian fighter jets conduct offensive action over Kosovo, but did in Libya (Husby 2015; Frost-Nielsen 2009)?

If this approach is inadequate, we may try a different one: we may look at what it is that impels small states to make use of coercive foreign policy instruments in remote locations in the relational logic between large and small alliance partners. Small state governments' struggle when the country reserves itself against invitations to intervene in crises, especially when the invitation comes from the country's principal ally and main guarantor of security (United States, NATO).

An argument can be made that the logic of power politics also applies *inside* security alliances. The message originating from the Melian Dialogue of the Peloponnesian War (460–446 BC), "The strong do as they can, and the weak suffer what they must" (Thucydides 1972: 406), runs through real political thinking and theory formation in IR like a red thread, from Thucydides, Machiavelli (1998 [1513]), and Hobbes (1973 [1651]), to Carr (1939), Morgenthau (1946), Waltz (1959, 1979), and Mearsheimer (2001). It is only by adapting to the premises laid down by the superior power that the survival of small states can be assured, and the small states' modest foreign policy ambitions can be sought.

This message, we may capture in yet another paraphrasing of Rufus Miles' Law: "*Where you stand, depends on who is asking.*" Applied to the study of the politics of caveats, this insight makes us expect that small-state alliance partners are particularly sensitive to the requests of alliance leaders relating to whether and how to contribute to coalition forces. The stronger, more demanding and specific the request is, the slimmer is the small state SPM for denying the wishes of the great power. This would likely include the issue of applying restrictive caveats on military contribution to coalition forces.

A final point relates to the ambition of constructing analytical tools capturing the dialectical relationship between inside-out and outside-in impulses to the foreign policy-making process. One such tool is Robert D. Putnam's (1988) "two-level games" analysis briefly mentioned above. A further tool is found in the literature on Transnational Relations (TR). The first studies on how multinational companies influence the global economic division of labor were conducted in the field of FPA. The research focused on the implications of the growth of transnational actors and networks for states' foreign policy SPM (Rosenau 1969; Nye and Keohane 1971; Vernon 1971; Mansbach 1976). The outside-in approach of TR has also been applied in more recent research (Risse-Kappen 1995; Josselin and Wallace 2002).

In the article "Second image reversed," Peter A. Gourevitch (1978) reverses the causal direction in Kenneth N. Waltz' "Man, State, War" (1959): instead of looking at how characteristics of states influence for-eign policy, and thus, in turn, the global environment (inside-out), Gourevitch looks at how transnational actors, markets, and structures not only influence states' foreign policy SPM and behaviour directly. Gourevitch is primarily preoccupied explaining how transnational net-works affect the states' foreign policy by the indirect route of changing the states' governance agencies and the power balance among domestic social groups. In a final rewording of Rufus Miles' Law, we may account for this transnational influence on foreign policy-making as follows:

"Where you stand, depends on your transboundary interests and prospects." This proposition inspires reasoning and research as to whether alliance members' policy toward participation and use of caveats in coalition forces can be related to states' need to secure markets and strategic resources in or related to the theatre of war.

The main gist of Gourevitch's explanatory thrust can be seen as a transnational parallel to Putnam's intergovernmental metaphor: While Putnam is concerned with how the domestic win-set and the foreign policy SPM together "determine" the governments' positions in international negotiations, Gourevitch sets out to trace the causal path from transnational actors and markets via national group interests to foreign policy behaviour. Both are relevant responses to Christopher Hill's research challenge to the FPA community:

> What is important [...] is to study the dialectical relationship between internal and external forces [concerning Foreign Policy], and to show how that is important for both "inside" and "outside". (Hill 1996: 9)

We could not agree more. In the FPA approach to the politics of caveats, we are invited study how domestic and global politics interplay in influencing the foreign policy decision-making processes that produce decisions on the participation of coalition forces. We are further encouraged to research how attributes of the state governmental apparatus and key decision-makers influence perceptions of SPM and preferences, and the choice of policy instruments, caveats included. In a fully integrated FPA model, all these levels of analyses are accounted for (Fermann 2013: 108–128).

In the present chapter, we have offered an introduction to different libraries of Political Science to indicate how they may apply in FPA of the politics of caveats. In Chapters 8–10 (Part 4), we engage in more focused and applied theorizing. Having contextualized and conceptualized the concept of caveats (Part 2), and explained the analytical framework and bodies of literature to tap from (Part 3), we now enter the phase of reasoning where we theorize specifically to generate hypotheses on conditions for coalition participation and the use of national reservations on the use of force.

REFERENCES

Adler, E. (1997). Seizing the Middle Ground: Constructivism in World Politics. *European Journal of International Relations, 3*(3), 319–363.

Adler, E. (2002). Constructivism and International Relations. In W. Carlsnaes, T. Risse, & B. Simmons (Eds.), *Handbook of International Relations* (pp. 95–118). London: Sage.

Allison, G. T. (1971). *Essence of Decision: Explaining the Cuban Missile Crisis.* Boston: Little, Brown.

Almond, G. A. (1950). *The American People and Foreign Policy.* New York, NY: Praeger Publishers.

Almond, G., & Verba, S. (1963). *The Civic Culture: Political Attitudes and Democracy in Five Nations.* Thousand Oaks, CA: Sage.

Beal, R. S., & Hinckley, R. H. (1984). Presidential Decision Making and Opinion Polls. *The ANNALS of the American Academy of Political and Social Science, 472,* 72–84.

Boynton, G. R. (1991). The Expertise of the Senate Foreign Relations Committee. In V. M. Hudson (Ed.), *Artificial Intelligence and International Politics* (pp. 291–309). Boulder, CO: Westview Press.

Breslauer, G. W., & Tetlock, P. (Eds.). (1991). *Learning in US and Soviet Policy.* Boulder, CO: Westview Press.

Bull, H. (1977). *The Anarchical Society: A Study of Order in World Politics.* New York, NY: Columbia University Press.

Bull, H., & Watson, A. (Eds.). (1984). *The Expansion of International Society.* Oxford: Clarendon Press.

Campbell, A. A., Converse, P. E. M., Miller, W. E., & Stokes, D. E. (1964). *The American Voter.* New York, NY: Wiley.

Carr, E. H. (1939). *The Twenty Years' Crisis.* London: Macmillan.

Clarke, M. (1996). Foreign Policy Analysis: A Theoretical Guide. In S. Stavridis and C. Hill (Eds.), *Domestic Sources of Foreign Policies. Western European Reactions to the Falkland Conflict* (pp. 19–39). Oxford: Berg.

Cottam, R. (1977). *Foreign Policy Motivation: A General Theory and a Case-Study.* Pittsburg, PA: University of Pittsburg Press.

Dahl, R. (Ed.). (1973). *Regimes and Oppositions.* New Haven: Yale University Press.

Dale, O.-M. (2005). *Den norske beslutningen om ikke å delta i invasjonen av Irak, men å delta i okkupasjonen av landet (2003).* Master Thesis in Political Science, Trondheim, Department of Sociology and Political Science, Norwegian University of Science and Technology (NTNU).

De Rivera, J. (1968). *The Psychological Dimension of Foreign Policy.* Columbus, OH: Merrill Publishing.

East, M. A., Salmore, S. A., & Hermann, C. F. (Eds.). (1978). *Why Nations Act. Theoretical Perspectives for Comparative Foreign Policy Studies.* Beverly Hills, CA: Sage.

Evans, P. B., Jacobson, H. K., & Putnam, R. (Eds.). (1993). *Double-Edged Diplomacy: International Bargaining and Domestic Politics.* Berkeley, CA: University of California Press.

Fermann, G. (1994). Retorikeren, analytikeren og den prinsipielle observatør: Maktbruk og berettigelse i Gulf-konflikten. *Norsk statsvitenskapelig tidsskrift, 10*(2), 291–309.

Fermann, G. (1997). Political Context of Climate Change. In G. Fermann (Ed.), *International Politics of Climate Change: Key Issues and Critical Actors* (pp. 9–46). Oslo, Oxford, Boston: Scandinavian University Press.

Fermann, G. (Ed.). (2013). *Utenrikspolitikk og norsk krisehåndtering.* Oslo: Cappelen Damm Akademika.

Frost-Nielsen, P. M. (2009). *Rules of Engagement. En utenrikspolitisk case-analyse av den politiske kontrollen av norske kampfly i Operation Enduring Freedom, Afghanistan 2002–2003.* Master Thesis in Political Science, Trondheim, Department of Sociology and Political Science, Norwegian University of Science and Technology (NTNU).

George, A. A. (1969). The "Operational Code": A Neglected Approach to the Study of Political Leaders and Decision-Making. *International Studies Quarterly, 13*(2), 190–222.

Gerner, D. J., Schrodt, P. A., Francisco, R. A., & Weddle, J. (1994). Machine Coding of Events Data Using Regional and International Sources. *International Studies Quarterly, 38*(1), 91–120.

Goffman, E. (1966). *The Presentation of Self in Everyday Life.* Edinburgh: University of Edinburgh, Social Sciences Research Centre.

Gourevitch, P. (1978). Second Image Reversed: The International Sources of Domestic Politics. *International Organization, 32*(4), 881–912.

Hagan, J. D. (1995). Domestic Political Explanations in the Analysis of Foreign Policy. In L. Neack, J. A. K. Hey, & P. J. Haney (Eds.), *Foreign Policy Analysis* (pp. 117–143). Englewood Cliffs, NJ: Prentice-Hall.

Halperin, M. H. (1974). *Bureaucratic Politics and Foreign Policy.* Washington, DC: Brookings Institution.

Hart, P. T., Stern, E. H., & Sundelius, B. (Eds.). (1997). *Beyond Groupthink: Political Group Dynamics and Foreign Policy-Making.* Ann Arbor, MI: University of Michigan Press.

Hermann, M. G. (1978). Effects of Personal Characteristics of Leaders on Foreign Policy. In M. A. East, S. A. Salmore, & C. F. Hermann (Eds.), *Why Nations Act. Theoretical Perspectives for Comparative Foreign Policy Studies* (pp. 44–68). Beverly Hills, CA: Sage.

Hermann, M. G., & Hermann, R. K. (1989). Who Makes Foreign Policy Decisions and How? *An Empirical Inquiry. International Studies Quarterly, 33*(4), 361–387.

Hermansson, H., & Fermann, G. (2013). Myndighetenes legitimering av norsk deltakelse I NATO-operasjoner I Bosnia, Kosovo og Afghanistan. In G. Fermann (Ed.), *Utenrikspolitikk og norsk krisehåndtering* (pp. 335–354). Oslo: Cappelen Akademisk Forlag.

Herrmann, R. K. (1985). *Perceptions and Behaviour in Soviet Foreign Policy.* Pittsburg, OH: University of Pittsburg Press.

Herrmann, R. K. (1993). *The Construction of Images in International Relations Theory: American, Russian and Islamic World Views.* Paper presented at the 34th Annual Conference of the International Studies Association.

Herrmann, R. K., & Kegley, C. W. (1995). Rethinking Democracy and Democratic Peace. Perspectives from Political Psychology. *International Studies Quarterly, 39*(4): 511–534.

Heuer, R. J. (1999/1978–86). *The Psychology of Intelligence Analysis.* Washington, DC: Government Printing Office.

Hill, C. (1996). Introduction: The Falklands War and European Foreign Policy. In S. Stavridis & C. Hill (Eds.), *Domestic Sources of Foreign Policy* (pp. 1–18). Oxford: Berg.

Hill, C. (2003). *The Changing Politics of Foreign Policy.* Houndmills, Basingstoke: Palgrave Macmillan.

Hilsman, R. (1967). *To Move a Nation. The Politics of Foreign Policy in the Administration of John F. Kennedy.* New York, NY: Doubleday.

Hobbes, T. (1973 [1651]). *Leviathan.* London: Dent. http://www.gutenberg.org/files/3207/3207-h/3207-h.htm.

Hopf, T. (2002). *Social Construction of International Politics: Identities and Foreign Policies, Moscow, 1955 and 1999.* Ithaca, NY: Cornell University Press.

Hudson, V. M. (2008). The History and Evolution of Foreign Policy Analysis. In S. Smith, A. Hadfield, & T. Dunne (Eds.), *Foreign Policy: Theories, Actors, Cases* (pp. 11–30). Oxford: Oxford University Press.

Huntington, S. P. (1960). Strategic Planning and the Political Process. *Foreign Affairs, 38*(2), 285–299.

Husby, G. (2015). *Fra hull i luften, til hull i Gaddafis bunker. Bruk av politiske reservasjoner på norsk militærmakt i flernasjonale koalisjonsoperasjoner. En komparativ studie av F-16 bidragene i Kosovo, Afghanistan og Libya.* Master Thesis in Political Science, Trondheim, Department of Sociology and Political Science, Norwegian University of Science and Technology (NTNU).

Janis, I. L. (1972). *Victims of Groupthink: A Psychological Study of Foreign-Policy Decisions and Fiascos.* Boston, MA: Mifflin.

Jervis, R. (1976). *Perception and Misperception in International Politics.* Princeton, NJ: Princeton University Press.

Johnson, L. K. (1977). Operational Codes and the Prediction of Leadership Behavior. In M. G. Hermann (Ed.), *A Psychological Examination of Political Leaders* (pp. 82–119). New York, NY: Free Press.

Johnston, A. I. (1995). Thinking About Strategic Culture. *International Security, 19*(4), 32–64.

Josselin, D., & Wallace, W. (Eds.). (2002). *Non-state Actors in World Politics.* Houndmills, Basingstoke: Palgrave.

Kaarbo, J. (2015). A Foreign Policy Analysis Perspective on Domestic Politics Turn in IR Theory. *International Studies Review, 17*(2), 189–216.

Katzenstein, P. J. (1985). *Small States in World Markets. Industrial Policy in Europe.* Ithaca, NY: Cornell University Press.

Kehr, E. (1973 [1927]). *Battleship Building and Party Politics in Germany 1894–1901: A Cross-Section of the Political, Social, and Ideological Preconditions of German Imperialism.* Chicago, IL: University of Chicago Press.

Keohane, R. O. (1989). *International Institutions and State Power: Essays in International Relations Theory.* Boulder, CO: Westview Press.

Keohane, R. O., & Nye, J. (1977). *Power and Interdependence: World Politics in Transition.* Boston, MA: Little, Brown.

Khanman, D. P., Slovic, P., & Tversky, A. (1982). *Judgment Under Uncertainty: Heuristic and Biases.* Cambridge: Cambridge University Press.

Khong, Y. F. (1992). *Analogies at War: Korean, Munich, Dien Bien Phu, and the Vietnam Decision of 1965.* Princeton, NJ: Princeton University Press.

Larson, D. W. (1985). *Origins of Containment: A Psychological Explanation.* Princeton, NJ: Princeton University Press.

Lebow, R. N., & Stein, J. G. (1990). Deterrence: The Elusive Dependent Variable. *World Politics, 42*(3), 336–369.

Levy, J. S. (2003). Political Psychology, and Foreign Policy. In D. O. Sears, L. Huddy, & R. Jervis (Eds.), *Oxford Handbook of Political Psychology* (pp. 253–284). New York, NY: Oxford University Press.

Lindblom, C. E. (1959). The Science of "Muddling Through". *Public Administration Review, 19*(1), 79–88.

Machiavelli, N. (1998 [1513]). *Fyrsten.* Oslo: Grøndahl & Dreyer.

Mansbach, R. W. (1976). *The Web of World Politics: Non-state Actors in the Global System.* London: Prentice Hall.

McGowan, P., & Shapiro, H. B. (1973). *The Comparative Study of Foreign Policy: A Survey of Scientific Findings.* Beverly Hills, CA: Sage.

Mearsheimer. J. J. (2001). *The Tragedy of Great Power Politics.* New York, NY: Norton & Co.

Milner, H. (1993). The Interaction of Domestic and International Politics: The Anglo-American Oil Negotiations and the International Civil Aviation Negotiations, 1943–1947. In P. B. Evans, H. K. Jacobson, & R. D. Putnam

(Eds.), *Double-Edged Diplomacy. International Bargaining and Domestic Politics* (pp. 207–232). Berkeley, CA: University of California Press.

Moravcsik, A. (1993). Introduction: Integrating International and Domestic Theories of International Bargaining. In P. B. Evans, H. K. Jacobson, & R. D. Putnam (Eds.), *International Bargaining and Domestic Politics. Double-Edged Diplomacy* (pp. 3–42). Berkeley, CA: University of California Press.

Moravcsik, A. (1997). Taking Preferences Seriously: A Liberal Theory of International Politics. *International Organization, 51*(4), 513–553.

Morgenthau, H. J. (1946). *Politics Among Nations. Scientific Man vs. Power Politics.* Chicago, IL: Chicago University Press.

Morse, E. (1976). *Modernization and the Transformation of International Relations.* New York, NY: The Free Press.

Nacos, B. L., Shapiro, R. Y., & Isernia, P. (Eds.). (2000). *Decision-Making in a Glasshouse. Mass Media, Public Opinion, and American and European Foreign Policy in the 21st Century.* New York, NY: Rowman & Littlefield.

Neumann, I. B. (2012). *At Home with the Diplomats: Inside a European Foreign Ministry.* Ithaca, NY: Cornell University Press.

Neustadt, R. E. (1970). *Alliance Politics.* New York, NY: Columbia University Press.

Nye, J. S., & Keohane, R. (Eds.). (1971). *Transnational Relations and World Politics.* Cambridge, MA: Harvard University Press.

O'Neill, T., & W. Novak. (1987). *Man of the House: The Life and Political Memoirs of Speaker Tip O'Neill.* London: Random House.

Onuf, N. G. (1989). *World of Our Making. Rules and Rule in Social Theory and International Relations.* Columbia, SC: University of South Carolina Press.

Paige, G. D. (1968). *The Korean Decision, June 24–30, 1950.* New York, NY: Free Press.

Putnam, R. D. (1988). Diplomacy and Domestic Politics. The Logic of Two-Level Games. *International Organization, 42*(3): 427–460.

Pye, L., & Verba, S. (1965). *Political Culture and Political Development.* Princeton, NJ: Princeton University Press.

Richardson, N. R., & Kegley, C. W. (1980). Trade Dependence and Foreign Policy Compliance: A Longitudinal Analysis. *International Studies Quarterly, 24*(2), 191–222.

Ringmar, E. (1996). *Identity, Interest, and Action.* Cultural Explanation of Sweden's Intervention in the Thirty Years War. Cambridge: Cambridge University Press.

Risse-Kappen, T. (Eds.). (1995). *Bringing Transnational Relations Back In: Non-state Actors, Domestic Structures, and International Institutions.* Cambridge: Cambridge University Press.

Rosenau, J. M. (1967). *Domestic Sources of Foreign Policy.* New York, NY: Free Press.

Rosenau, J. M. (1969). *Linkage Politics.* New York, NY: Free Press.

Rummel, R. J. (1979). *National Attitudes and Behaviors.* Beverly Hills, CA: Sage.

Russet, B., & Shye, S. (1993). Aggressiveness, Involvement, and Commitment to Foreign Policy Attitudes: Multidimensional Scaling. In D. Caldwell & T. McKeown (Eds.), *Diplomacy, Force, and Leadership: Essays in Honor of Alexander L. George.* Boulder, CO: Westview Press.

Schilling, W. R., Hammond, P. Y., & Snyder, G. H. (1962). *Strategy, Politics, and Defense Budgets.* New York, NY: Columbia University Press.

Schrodt, P. A. (1995). Event Data in Foreign Policy Analysis. In L. Neack, J. A. Hey, & P. J. Haney (Eds.), *Foreign Policy Analysis: Continuity and Change in Its Second Generation* (pp. 145–166). Englewood Cliffs, NJ: Prentice-Hall.

Selb, P. (1996). *Headline Diplomacy: How News Coverage Affects Foreign Policy.* London: Praeger Publishers.

Shapiro, M. J., & Bonham, M. (1973). Cognitive Process and Foreign Policy Decision-Making. *International Studies Quarterly, 17*(2), 147–174.

Simon, H. A. (1947). *Administrative Behavior.* New York, NY: Free Press.

Simon, H. A. (1957). *Models of Man: Social and Rational.* New York, NY: Wiley.

Simon, H. A. (1985). Human Nature in Politics: The Dialogue of Psychology with Political Science. *American Political Science Review, 79*(2), 293–304.

Singer, E., & Hudson, V. M. (Eds.). (1992). *Political Psychology and Foreign Policy.* Boulder, CO: Westview Press.

Skidmore, D., & Hudson, V. M. (1993). *The Limits of State Autonomy: Societal Groups and Foreign Policy Formulation.* Boulder, CO: Westview Press.

Snyder, R. C., Bruck, H. W., & Sapin, B. (Eds.). (1954). *Decision-Making as an Approach to the Study of International Politics. Foreign Policy Analysis Project Series No. 3.* Princeton, NJ: Princeton University Press.

Sprout, H., & Sprout, M. (1956). *Man-Mileu Relationship Hypotheses in the Context of International Politics.* Princeton, NJ: Princeton University Press.

Sprout, H., & Sprout, M. (1957). Environment Factors in the Study of International Politics. *Journal of Conflict Resolution, 1*(3), 309–328.

Sprout, H., & Sprout, M. (1965). *The Ecological Perspective on Human Affairs with Special Reference to International Politics.* Princeton, NJ: Princeton University Press.

Stavridis, S., & Hill, C. (Eds.) (1996). *Domestic Sources of Foreign Policy.* Oxford: Berg.

Steinbruner, J. D. (1974). *The Cybernetic Theory of Decision.* Princeton, NJ: Princeton University Press.

Stillman, R. (1999). "Where You Stand Depends on Where You Sit" (or, Yes, Miles's Law also Applies to Public Administration Basic Texts). *American Review of Public Administration, 29*(1), 92–97.

Thucydides. (1972). *History of the Peloponnesian War.* London: Penguin Books.

Vaara, E., & Tienari, J. (2001). Justification, Legitimization, and Naturalization of Mergers and Acquisitions. *A Critical Discourse Analysis of Media Texts. Organization, 9*(2): 275–304.

Vernon, R. (1971). *Sovereignty at Bay. The Multinational Spread of US Enterprises.* New York, NY: Basic Books.

Walker, R. B. J. (1993). *Inside/Outside: International Relations as Political Theory.* Cambridge: Cambridge University Press.

Walt, S. (1992). Revolution and War. *World Politics, 44*(3), 321–368.

Waltz, K. N. (1959). *Man, State, War.* New York, NY: Columbia University Press.

Waltz, K. N. (1979). *Theory of International Politics.* Reading: Addison-Wesley.

Weber, M. (1995 [1920]). *Makt og byråkrati. Essays om politikk og klasse, samfunnsforskning og verdier.* Oslo: Gyldendal (in Norwegian).

Wendt, A. (1994). Collective Identity Formation and the International State. *American Political Science Review, 88*(2), 384–396.

Wendt, A. (1999). *Social Theory of International Politics.* Cambridge: Cambridge University Press.

PART IV

Theorizing Caveats

CHAPTER 8

Alliance Politics Dynamics

In the introductory chapter, we speculated whether caveats may limit the political costs of participating in coalition forces motivated by alliance obligations. In the present chapter, we elaborate on a theoretical argument as to how alliance politics may influence foreign policy-makers' decisions on coalition participation and use of caveats. Before traveling this deductive route, recall that caveats, the phenomenon under scrutiny was defined as national reservations (restrictive or permissive) on the use of force in a coalition context. Such reservation is politically motivated and should be distinguished from restrictive behavior that is due to coordination failure or lack of capacity. In Chapter 4 it was concluded that the application of caveats is observed in national deviations from coalition RoE, in national interference in the coalition's chain of command, and in national limitation on the extent to which coalition is delegated the authority to make full use of the operational capacity of the national contingent as to where, when and how the contingent be deployed and used in theater of war.

Recall also from Chapter 2 the contextualization of caveats as the final step in a foreign policy-making process that reveals how enthusiastic, lukewarm, or skeptical the potential coalition member is toward the military mission in question: First is the decision whether to participate in the coalition. Second, if participating, should the contribution be a symbolic display of political support, or a substantial military contingent? Third, if agreeing on a substantial military contribution, shall any national reservations on the use of force apply? Finally, if so, what kind and extent of caveats would serve the contributing government's principal political purpose for participating in the coalition?

© The Author(s) 2019   127
G. Fermann, *Coping with Caveats in Coalition Warfare*,
https://doi.org/10.1007/978-3-319-92519-6_8

In each step of the decision-making process the potential coalition-contributing government adjusts and fine-tune its response to coalition invitation to balance concerns about alliance commitments, interests in the outcome, and international norms regulating matters of war. Even before having decided to commit substantial military contingents to the coalition, coalition members have strong incentive to start worrying about the risks, and the political and material costs of coalition participation.

Arguably, states' cost-benefit calculations on their mission-related security interests and the legal status of the military intervention (Jus ad Bellum) may explain the decision to participate in coalitions. However, further considerations are required to explain variations in how states contribute to coalition forces, and in particular why some governments attach caveats to their military contribution. This distinction resonates with David P. Auerswald and Stephen M. Saideman's observation that states' use of caveats in the ISAF-campaign in Afghanistan did not depend on the degree to which states were affected by acts of international terrorism (2014: 16–19). Hence, Auerswald and Saideman turn their attention to domestic sources of explanation of the pattern of caveats in ISAF. In Chapters 9 and 10 we respond to this invitation by discussing how caveats are explained by governmental attributes, and the need to politically control military implementation in the theater of war. In the present chapter, we discuss how certain aspects of global politics nevertheless may influence states propensity to apply caveats in coalition operations (Frost-Nielsen 2016: 24–29; 2017: 3–6).

The balancing of security threats in international politics is central to Auerswald and Saideman's research and the reason why they research the degree to which coalition-contributing states in ISAF were affected by acts of international terrorism. However, balancing security threats is not the only possible systemic level explanation of caveats. Research has shown that under various conditions, other systemic explanations may account for states' level of support in military coalitions (Baltrusaitis 2010; Bennett et al. 1997; Davidson 2011). Without further testing of cases other than NATO's ISAF-campaign, it is premature to discard the impact of external factors on how states' choose to support coalitions. One explanatory possibility within the tenets of Political Realism is that caveats are the outcome of how states balance their alliance commitments.

Glenn H. Snyder (1997: 165–166) argues that states entering into an alliance[1] share common interests in preserving the alliance for the joint benefit of strengthened security. Still, divergent and conflicting interests remain, which variably threaten to pull allies apart. Allies constantly bargain with each other to counter these centrifugal forces to maximize joint benefits and minimize costs. Coordination of military plans in particular coalition operations is one of the most prominent issues over which such intra-alliance bargaining takes place.

Snyder offers clues to understand the finer grain of *how* states choose to participate in coalition forces. In his work on the security dilemma in alliance politics (1984, 1997), Snyder develops an analytical tool to explain caveats as the outcome of the decisions of hesitant allies that have few if any particular interests in favor of partaking in the mission, except to honor the commitment to support allies when collective threats to security arise.

The honoring of commitments in international politics is hardly an act rooted in altruism, but rather investments in long-term self-interest. By accepting the commitment to offer military assistance, a reluctant government invests in the continuing credibility of the solidarity norm, which is at the core of any collective security arrangement based on consent. What on the surface looks like an interest-based investment in the political objectives of the particular mission, may rather be motivated as an installment in security insurance with the expectation that it will pay back when the table turns sometime in the future. Still, long-term investment in the alliance and short-term mission-related risks, costs, and benefits need balancing for any coalition invitee. While participation in the coalition is an investment in the long-term security arrangement that is the alliance, the application of caveats may be interpreted as a concession to more immediate risks and costs related to the participation.

Thus, the foreign policy instrument of caveats may make it politically feasible for a reluctant government to participate militarily in a coalition force were few if any short-term individual gains are expected.

---

[1]Glenn H. Snyder treats security alignments and alliances as identical phenomena: "Alignments, whether or not they have been formalized as alliances, are essentially expectations in the minds of statesmen about whether they will be supported, opposed, or ignored by other states in future interactions" (1997: 21–22). It is reasonable to assume that Snyder's theoretical propositions apply to all kinds of security cooperation, from formal alliances to ad hoc "coalitions-of-the-willing."

The crucial implication of this line of reasoning is that in the improbable case caveats be prohibited as a legitimate policy instrument in alliance politics, reluctant coalition members may find that the more burdensome coalition participation outweigh the future expected utility of current investment in alliance solidarity. In such a circumstance, the reluctant coalition member may decide to fail on its alliance commitments by abandoning the coalition or restrict participation to merely a token contribution. In such a perspective, caveats are the lesser evil for the coalition and the best balancing of reluctant coalition members' long-term and short-term cost-benefit considerations.

In the final section, we deduce three empirical propositions on participation in coalition forces from Snyder's reasoning on the security dilemma in alliance politics. Before this, the burden-sharing scholarly debate is briefly reviewed, and it is explained how a selection of studies on alliance politics may contribute to our understanding of caveats in coalition operations.

BURDEN-SHARING DEBATES DURING AND AFTER THE COLD WAR

In alliance politics, external impulses to the foreign policy-making process may come in the shape of a request to buy into a new weapon platform (Aaberg 2016), as demands to increase the allocations to defense (Defense News 2017), or as an invitation to participate in a coalition force. The call to participate in the coalition forces in the 1999 Kosovo conflict and the 2011 Libya crisis were hard to ignore for several member states of NATO. This was, even more, the case after the 2001 November 11 attacks on the United States. Invoking Article 5 of the North Atlantic Treaty for the first time since the establishment of NATO in 1949,[2] the United States called upon allies to support military operations in Afghanistan. Throughout the 2000s, NATO members confirmed their decisions to contribute to the coalition operations in this distant theater of war at regular intervals. However, not without hesitation.

To understand better what decision-making problems the foreign policy instrument of caveats may solve, we may learn from the study of

---

[2]The principle of collective defense is enshrined in Article 5 of the Washington Treaty, which is the formal name for the North Atlantic Treaty underpinning NATO. Collective defense implies that an attack against one ally treated as an attack against all allies.

the dynamics of alliance politics. Most explicitly, national reservations on the use of force discussed in the literature on burden sharing in military alliances. This body of research originates from the debate on collective action problems related to the economic burden sharing between NATO member-states during the Cold War (Murdoch and Sandler 1991).

The main argument from the debate is that the alliance leader (a great power) makes the chief contribution to upholding the collective good of defense and security. This is partly because the alliance leader has the superior capacity to do so. But also because the alliance leader needs the alliance to balance the perceived threat from another great power with an alliance system of its own. In this way, the alliance leader is structurally induced to invest disproportionately in an alliance system that provides a public good (increased collective security) that cannot be withheld from any member of the alliance without considerable loss of credibility. Sensing the alliance leader's strong motivation for preserving the credibility and cohesion of the alliance, the weaker members of the alliance have a free-rider incentive to contribute less than their "fair share" to the security arrangement (Olson and Zeckhauser 1966).

During the Cold War, several contributors to this literature were engaged in measuring the variation in burden sharing among NATO member states by comparing annual military expenditures. The research borrowed insights from economic theory and results from the research were published in economic journals. Insights from this formalized literature were transferred to the analysis of new forms of burden sharing after the Cold War (Chalmers 2001).

With the demise of the Soviet Union, scholarship refocused on new security challenges and tasks for NATO. It was no longer as relevant to assess burden sharing and problems of free riding in terms of military expenditures. Instead, the question of burden sharing now related to the member states' political will and military capacity to make highly qualified military assets available for NATO's coalition operations far beyond the security organization's traditional area of operation. Caveats became an ingredient in this new burden-sharing security mosaic. National reservations on the use of force were considered one of the several novel expressions of degree of willingness to contribute in the new NATO (see, e.g., Cimbala and Forster 2010; Hallams and Schreer 2012; Lepgold 1998; Matlary and Petersson 2013; Noetzel and Schreer 2009; Ringsmose 2010; Sandler and Shimizu 2014). The second-generation scholarly debate on burden sharing in NATO had the potential to shed light on the politics of caveats. However, it did so only to a limited extent by choosing to study national reservations on the use of force merely as another problem that needs to be solved in allied burden-sharing negotiations.

The reason for this limited focus is probably that many scholarly contributions build on the ideal assumption that military alliances and coalitions are merely mechanisms for the mobilization of military strength between states with overlapping or complementary security interests. In such analyses, the premise is that skewed burden sharing and free riding is a problem which leads to underproduction of military striking-power and collective security, thus weakening and possibly eroding the political rationale for the alliance. Within this approach, caveats and other forms of reserved alliance behavior are understood as some deviation from an ideal state of cooperation among states with common security interests (Crawford 2014; Driver 2016; Hartley and Sandler 1999; Overhage 2013; Richter and Webb 2014).

This line of reasoning led scholars to study caveats as a phenomenon potentially undermining alliance solidarity, and is less capable of acknowledging how national reservations on the use of force may become the lesser evil to the alternative of desisting from participating in the coalition altogether. Failing to consider how caveats may reduce the political costs of honoring alliance commitments obscures the potential windfall from applying national reservations on the use of force in coalition contexts. The FPA approach induces us to refocus from seeing caveats merely as a threat to effective collective action. In FPA, we study caveats as a problem-solving instrument in foreign policy-making, and as a political contribution to the fine-tuning and optimization of the trade-offs which needs to be made in any burden-sharing negotiation. There is thus a demand for research that can shed light on how the application of national reservations on the use of force makes it politically feasible for governments to answer in the affirmative invitations to contribute militarily to coalition forces.

CAVEATS AS A MEANS TO BALANCE THE SECURITY DILEMMA IN ALLIANCE POLITICS

To understand how alliance dynamics may create a demand for caveats, we need to look at other contributions to the study of alliance politics. Glenn H. Snyder (1984, 1997) has not explicitly studied caveats but has developed insights that can shed light on conditions and mechanisms capable of explaining some of the empirical variations in the use of caveats across military contributions and coalition forces. Snyder has shown that while alliances is a mechanism for the mobilization of military capacities, states' alliance behavior is due to more complex reasons than the simplified assumption that alliance politics is merely about force generation.

Snyder assumes that states have two motives for joining an alliance. One is to strengthen own security by deriving benefit from other states' contributions to the military alliance. The other motive is to maximize own utility of belonging to the alliance. On the one hand, states attempt to gain as much security as possible from the commitments and security contributions of other states. On the other hand, states' are inclined to limit own commitment to defend other alliance members' security interests which do not correspond with own individual self-interests.

Second, Snyder argues that the collective security concerns inducing states to organize in security alliances cannot explain alliance burden-sharing behavior *after* the establishment of the alliance. In managing alliance politics, every member of the security arrangement has to cope with the possibility that other states are less committed to the alliance. Recall that the political rationale for establishing the alliance is the mutual expectation that the coordination efforts will increase the security of each one, and all of the member states. However, a state strongly committed to the alliance in words and deeds may nevertheless have reason to fear being drawn into risky military conflicts were few if any national interests served other than the potential future benefits of demonstrating alliance solidarity. The committed alliance member may also fear that other states are not fully committed to the alliance and will fail to act on their commitments when the alliance and other member states in dire straits. In that case, the added security the alliance established to provide is a mirage.

Snyder's line of reasoning leads to the security dilemma in alliance politics in the purest form: A member state's strong commitment to the collective security system of the alliance is an investment in the cohesion and future existence of the alliance. However, solidary behavior increases the danger of being tied down in dangerous and costly conflicts with no windfall. These are the realities of member state's *fear of entrapment*. However, this is not the alliance member's only concern. The mirror image of the fear of entrapment is the *fear of abandonment*. The two are functionally linked: A weaker commitment to the alliance implies less danger of being caught up in conflicts in which the alliance member has no positive interest invested, but increases the risk of alliance partners reciprocating the lukewarm and minimal investment in alliance solidarity at a later occasion. In the worst of cases, reserved alliance behavior may become contagious if not checked by the alliance leader, and spiral into member states jumping ship, and the subsequent severe weakening of the alliance.

A member state's commitment to the alliance is thus a complex function of the state's consideration of costs and risks related to loss of autonomy on the one hand, and particular gains and the expectation that other members of the alliance are willing to offer assistance when a threating situation arise in the future on the other (Morrow 1991). According to Snyder, it is precisely because a trade-off needs to be made between the dual fears of abandonment and entrapment that alliance members find themselves in a fundamental state of intra-alliance strategic ambiguity. This duplicity in how far one is to commit to the security alliance is the proverbial "elephant in the room" in ongoing deliberations as to how the alliance is to prioritize collective goals and resources (Davidson 2011; Kim 2011; Press-Barnathan 2006).

Snyder (1997: 166) argues that it is particularly in the coordination of military planning among military alliance partners that disagreements and opposing views are likely to surface. The policy instrument of national reservations on the use of force is one way of balancing and expressing strategic ambivalence on this negotiating arena. The decision to contribute forces to the coalition operation may reasonably be interpreted as a sign of commitment to the alliance due to fear of abandonment, or in support of the particular political objectives of the military mission. On the other hand, the application of caveats on the use of military contribution may be understood as a balancing concession to the fear of entrapment, the fear that an unreserved military contribution may inflict considerable military and political costs further down the road.

In this reading, the widespread use of national reservations on the use of force in NATO's ISAF campaign in Afghanistan (Auerswald and Saideman 2014; NATO 2005) are the expression of several coalition members making trade-offs between fear of abandonment and fear of entrapment. By invoking Article 5 of The North Atlantic Charter, the United States did not give lukewarm members of NATO any uncostly option to refuse to participate in the coalition. Even though all members of NATO (and beyond) publicly declared that it was in their interest to make sure Afghanistan did not become a haven where al-Qaida could operate with impunity and plan new attacks on Western states, disagreement within the alliance persisted as to how NATO should achieve this primary political objective operationally.

The main fracture was between member states in favor of a rather defensive military campaign that emphasized the "reconstruction of Afghanistan,", and alliance partners (the US, in particular) arguing that priority be granted to offensive "counter-insurgency" operations designed to break the Afghan insurgent's capacity to carry out extensive military operations. This kind of tactical disagreement led to continuous negotiations among the coalition partners as to how and where NATO should apply force to reach the political objectives of the campaign (Rynning 2012).

National reservations on the use of force provide a potential solution for states to manage this kind of disagreement. The contribution of military contingents to the Afghanistan campaign was an expression of alliance solidarity and a willingness to stand by security commitments. At the same time, it was an investment in security insurance, to reduce the risk of abandonment when the need for alliance support arise in the future. However, several contributing states attached restrictive reservations on the use of the contingents by, for instance, not allowing the coalition Force Commander to deploy the contingent when and where he saw fit. These kinds of restrictions made it almost a certainty that the contingent in question kept out of harm's way. For the Force Commander who was precluded from using some contingents effectively, it was the more crucial that the severely restricted contingents were not allowed to stand in the way of those contingents willing to fight the war where it was at the most demanding.

For our research purposes, we draw attention to the argument that restrictive caveats imposed on several contingents were implemented because of fear of being entrapped in the war in a capacity inconsistent with the national interest. The decision to contribute forces to the coalition war-effort is a display of alliance solidarity and an investment in abandonment avoidance. The decision to apply restrictive national reservations on the use of force is an indication that there are limits to alliance solidarity, and that the limit likely shows when policy-makers consider that concerns about entrapment outweighs concerns about abandonment.

Other Considerations: Historical Legacy, Force Integration, Heterogeneity

Glenn H. Snyder has provided crucial contributions to our understanding of how the security dilemma in alliance politics explain states' ambiguous behavior in alliances (1984, 1997). We have argued that restrictive caveats are an expression of coalition members balancing of conflicting concerns. However, Snyder himself does not say anything about the actual mechanisms through which caveats may contribute to strike politically feasible trade-offs between diverging concerns in alliance politics. Neither does he account for how different characteristics of alliances and coalitions may influence the security dilemma, and thus states' inclination to apply caveats to cope with it. On these questions, Patricia A. Weitsman's research provides more clues.

Through a series of publications, Weitsman has studied alliance behavior (Schneider and Weitsman 1997; Weitsman 1997, 2003, 2004, 2010, 2014). For one, she questions the assumption that alliances merely are mechanisms for the pooling of military power for states with common security interests. Weitsman argues that the establishment of alliances also are motivated as an institutional mechanism to bring about reconciliation between former adversaries. She makes a convincing argument that NATO was established as much to bring about the consolidation and reconciliation of Western European states recently locked into destructive relationships of animosity, as it was created to balance the Soviet threat to Western security (Weitsman 2010). Arguably, this dual motivation behind the establishment of the security organization is likely to amplify the security dilemma in NATO. If caveats are part of the solution to a security dilemma, the policy instrument should be in even greater demand if the competing fears of under- and overcommitting amplified by historically conditioned intra-alliance tensions and distrust.

Alliances and coalitions represent both possibilities and limitations on a state's ability to pursue own interests and is thus a two-edged sword in the pursuit of increased military power and security. Adding to coalition members' fear of overcommitting is the fact that alliances and coalitions accumulate military power not only by the amassment of national battalions, brigades, and divisions beside one another. To effectively utilize military resources, national contingents are integrated into a common military structure of command. If not disrupted, unified command allows for unambiguous command chains and swift military decision-making. However, the preservation of an egalitarian, consensus-based and time-consuming decision-making process will, in the midst of battle come at the expense of effective military leadership and use of military resources.

> To preserve sovereignty and serve their particular national interests, governments may intervene in the unified military command of coalition forces to retain some element of decision-making authority over their contingents. Such assertive arrangements can run the spectrum from insisting that decisions be made by consensus, to allowing troops to operate under the command of another state while still observing their national rules of engagement. (Weitsman 2014: 39)

Another means of preserving national control over national contingent in a highly integrated coalition force relates to the delegative dimension of RoE and caveats accounted for in Chapters 3 and 4 respectively. In this context, national reservations on the use of force is a solution to the coalition member's need to control how their forces are utilized to secure a reasonable balancing of the security dilemma in alliance politics. The red flag-holder in charge of saying "yes" or "no" to the Force Commander's request for use of the national contingent, based on instructions from the national government, is put in place to secure that the security dilemma of the contributing state is properly managed on a daily basis. An interesting question is whether the gate-keeping function of the red flag-holder or some national staff officer in the coalition chain of command are influenced (increased, diminished) by the degree of military integration in coalition force.

There is quite some variation as to what motivates and how motivated governments are within their decision to provide military contingents to coalition forces (Rapport 2015). Everything else being equal, a large coalition following benefits the cause of military force generation. However, the increasing size of the coalition is likely to come at the expense of added membership heterogeneity as to enthusiasm, motives, military equipment, and doctrines. Heterogeneity grows coordination problems that need to be solved at the expense of Force Commander's ability to effectively utilize the military capacity (Weitsman 2014). For such reasons, we may expect that the larger and more heterogeneous the coalition, the larger is the share of coalition members' applying caveats.

Sarah Kreps (2007, 2008) builds upon a similar argument. She finds that the United States choose to operate unilaterally in instances where the US government considers that highly efficient military action is required. Multilateral military operations are preferred when the need for broad political legitimacy overrides purely military and operational considerations. In exchange for legitimacy-demonstrating military support, the United States has been rather lenient toward coalition partners' application of national reservations on the use of force (Kreps 2008: 562; see also Weitsman 2014: 48–73). Related to the application of caveats is also the fact that NATO has developed decision-making institutions that have made the organization more capable of managing and coordinating national inputs, and thus counter the potential disruptive effect of caveats on the cohesion and military effectiveness of the coalition (Dijkstra 2015; Kupchan 1988; Reveron 2002).

Caveats Relevant Take-Home Messages

As to the relevance of caveats in alliance politics, the main take-home messages from our review of relevant scholarships are as follows: Restrictive caveats signals reluctant and lukewarm participation in coalition force. Generically speaking, national reservations on the use of force in coalition contexts indicate there are limits to common interests and alliance solidarity. Caveats are not itself the cause of restricted cooperation, but a symptom. At its most fundamental, national reservations on the use of force in coalition forces is a foreign policy instrument facilitating the balancing of the preservation of sovereignty and national control with the inclination to seek increased security through the membership in military alliance.

Caveats is yet another means of coping with and balance the security dilemma in alliance politics related to the dual fears of abandonment and entrapment. Within the premises of the literature on alliance politics, we argue that the application of caveats makes possible military contributions to coalitions which otherwise would not materialize (Frost-Nielsen 2016: 24–29). The instrument of caveats allows for compromises beyond the "all or nothing"-choice (Porter 2012). We may stretch the argument even further: Alliances cannot last without some allowance for the application of caveats. Restrictive caveats is a much less crude expression of reluctant participation than the alternatives of downsizing contribution or defecting from coalition force.

The alternative to the tensions caveats-use inflict in burden-sharing negotiations, in the chain of command, and in the field may thus be the failure to participate in the coalition in any capacity. Such uncompromising and free riding behavior has detrimental military and political consequences for the security alliance. In this line of reasoning, national reservations on the use of force is a foreign policy instrument and a means of statecraft capable of balancing different and only partly compatible concerns related to complex dynamics and dilemmas of alliance politics.

In studying military participation in coalition operations, the question should not be reduced to one in which the foreign policy-making state has an unequivocal interest to participate or not. Precisely because different forms of warfare serve different parties and concerns in different ways, the question of how an alliance partner is to participate in a coalition force need to be scrutinized with this particular state's collective and individual self-interests in mind. National reservations on the use of force is an instrument to strike a balance between diverging national concerns, and thereby also a tool to maximize coalition force generation within what is politically feasible.

POINTING THE ARGUMENTS: EMPIRICAL PROPOSITIONS

In the introductory chapter, we inquired about what decision-making problems and opportunities the application of national reservations on the use of force might address for foreign policy decision-makers contemplating to contribute a military contingent to some multinational military operation. At the level of global politics, we have argued that caveats limit the political costs of participating in coalition forces triggered mainly by alliance commitments. Instead of buck-passing entirely, caveats allows the lukewarm or hesitant coalition member to contribute the smaller proverbial buck.

When suspecting or identifying caveats at work in alliance politics, we may at this particular level of analysis ask *if* the empirical pattern of national reservations on the use of force is related to dynamics in alliance politics, and *how*. Having grasped important mechanisms at work, the final task is to tailor-make the formulation of empirical propositions and the operationalization of independent variables to particular research purposes. Three empirical propositions stand out for consideration as a point of departure for future empirical research:

"Alliance Politics" hypotheses (AP-H1–3): Everything else being equal, we may consider that;

AP-H1   *"Abandonment" hypothesis:* States primarily concerned with the risk of abandonment are inclined to provide unconditional military support to the coalition (no caveats).

AP-H2   *"Entrapment" hypothesis:* States primarily concerned with the risk of entrapment are not inclined to provide military support to the coalition (no participation whatsoever, caveats irrelevant).

AP-H3   *"Optimal trade-off" hypothesis:* States primarily concerned with the balancing of the dual risks of abandonment and entrapment are inclined to apply caveats to an extent compatible with the relative importance of the two concerns (measured application of caveats).

Empirical evidence for these basic propositions may come from public statements by executive officials such the prime minister, the foreign minister, or the minister of defense justifying participation in terms of alliance commitments (fear of abandonment), "while at the same time acknowledging risks and costs of such military participation" (fear of entrapment) (Frost-Nielsen 2017: 7). The extent of alliance commitment also shows in the material (in) action. Concerns about abandonment are expressed in a considerable military contribution to the coalition effort. Fear of entrapment shows in concerns of loss of own forces and collateral damage—the unintentional killing of civilians (ibid.). Empirical proposition AP-H3 is supported to the extent it can be found that caveats are put in place that prevents national contingents from taking part in risky and potentially costly military missions and tasks.

How may theorizing on the politics of caveats at the global level of analysis brought beyond the dynamics of alliance politics discussed in the previous? Where are other promising avenues of reasoning to be found? One answer is to take a sharper look at the distinction made between balance-of-threat and balance-of-power in the Structural Realist debate on defensive and offensive Realism (Waltz 1979; Walt 1987; Taliaferro 2001/2002; Mearsheimer 2001). Another avenue to make sense of the politics of caveats is to apply the literature on Political Economy, strategic interests, and strategic resources to understand better how the pecuniary and strategic interests in the area of deployment may influence coalition-behavior and policies on reservations on the use of force (Gilpin 2001; Klare 2009; Fermann 2014).

REFERENCES

Aaberg, M. (2016). *Kampflykjøp mellom barken og veden. En utenrikspolitisk analyse av beslutningen om å velge F-35 som Norges neste kampflyplattform.* Master thesis in Political Science, Department of Sociology and Political Science, Norwegian University of Science and Technology (NTNU), Trondheim.

Auerswald, D. P., & Saideman, S. M. (2014). *NATO in Afghanistan: Fighting Together, Fighting Alone.* Princeton, NJ: Princeton University Press.

Baltrusaitis, D. F. (2010). *Coalition Politics and the Iraq War.* Boulder, CO: First Forum Press.

Bennett, A., Lepgold, J., & Unger, D. (Eds.). (1997). *Friends in Need—Burden Sharing in the Persian Gulf.* New York, NY: St. Martin's Press.

Chalmers, M. (2001). The Atlantic Burden-Sharing Debate—Widening or Fragmenting? *International Affairs, 77*(3), 569–585.

Cimbala, S. J., & Forster, P. K. (2010). *Multinational Military Intervention: NATO Policy and Burden-Sharing.* Farnham: Ashgate.

Crawford, T. W. (2014). The Alliance Politics of Concerted Accommodation: Entente Bargaining and Italian and Ottoman Interventions in the First World War. *Security Studies, 23*(1), 13–147.

Davidson, J. W. (2011). *America's Allies and War: Kosovo, Afghanistan, and Iraq.* New York, NY: Palgrave Macmillan.

Defense News. (2017). *Trump's NATO Burden Sharing Fervor: Take It for a Drive at Munich.* http://www.defensenews.com/articles/trumps-natoburden-sharing-fervor-take-it-for-a-drive-at-munich.

Dijkstra, H. (2015). Functionalism, Multiple Principals and the Reform of the NATO Secretariat After the Cold War. *Cooperation and Conflict, 50*(1), 128–145.

Driver, D. (2016). Burden Sharing and the Future of NATO: Wandering Between Two Worlds. *Defense and Security Analysis, 32*(1), 4–18.

Fermann, G. (2014). What Is Strategic About Energy? De-simplifying Energy Security. In E. Moe & P. Midford (Eds.), *The Political Economy of Renewable Energy and Energy Security* (pp. 21–45). Basingstoke: Palgrave Macmillan.

Frost-Nielsen, P. M. (2016). *Betingede forpliktelser. Nasjonale reservasjoner i militære koalisjonsoperasjoner.* Ph.D. Dissertation in Political Science, Department of Sociology and Political Science, Norwegian University of Science and Technology (NTNU), Trondheim.

Frost-Nielsen, P. M. (2017). Conditional Commitments: Why States Use Caveats to Reserve Their Efforts in Military Coalition Operations. *Contemporary Security Policy, 38*(3), 371–397.

Gilpin, R. (2001). *Global Political Economy. Understanding the International Economic Order.* Princeton, NJ: Princeton University Press.

Hallams, E., & Schreer, B. (2012). Towards a 'Post-American' Alliance? NATO Burden Sharing After Libya. *International Affairs, 88*(2), 313–327.

Hartley, K., & Sandler, T. (1999). NATO Burden-Sharing: Past and Future. *Journal of Peace Research, 36*(6): 665–680.

Kim, T. (2011). Why Alliances Entangle but Seldom Entrap States. *Security Studies, 20*(3), 350–377.

Klare, M. T. (2009). *Rising Powers, Shrinking Planet: The New Geopolitics of Energy*. New York, NY: Holt and Company.

Kreps, S. E. (2007). The 1994 Haiti Intervention: A Unilateral Operation in Multilateral Clothes. *Journal of Strategic Studies, 30*(3): 449–474.

Kreps, S. E. (2008). When Does the Mission Determine the Coalition? The Logic of Multilateral Intervention and the Case of Afghanistan. *Security Studies, 17*(3): 531–567.

Kupchan, C. A. (1988). NATO and the Persian Gulf: Examining Intra-alliance Behavior. *International Organization, 42*(2): 317–346.

Lepgold, J. (1998). NATO's Post-Cold War Collective Action Problem. *International Security, 23*(1): 78–106.

Matlary, J. H., & Petersson, M. (Eds.). (2013). *NATO's European Allies Military Capability and Political Will*. New York, NY: Palgrave Macmillan.

Mearsheimer, J. (2001). *Tragedy of Great Power Politics*. New York, NY: W.W. Norton.

Morrow, J. D. (1991). Alliances and Asymmetry: An Alternative to the Capability Aggregation Model of Alliances. *American Journal of Political Science, 35*(4): 904–933.

Murdoch, J. C., & Sandler, T. (1991). NATO Burden Sharing and the Forces of Change: Further Observation. *International Studies Quarterly, 35*(1): 109–114.

NATO. (2005). *Resolution 336 on Reducing National Caveats*. Copenhagen. http://www.nato-pa.int/default.asp?SHORTCUT=828.

Noetzel, T., & Schreer, B. (2009). Does a Multi-tier NATO Matter? The Atlantic Alliance and the Process of Strategic Change. *International Affairs, 85*(2), 211–226.

Olson, M., & Zeckhauser, R. (1966). An Economic Theory of Alliances. *Review of Economics and Statistics, 48*(3), 266–279.

Overhage, T. (2013). Pool It, Share It, or Lose It: An Economical View on Pooling and Sharing of European Military Capabilities. *Defence & Security Analysis, 29*(4), 323–341.

Porter, P. (2012). A Matter of Choice: Strategy and Discretion in the Shadow of World War II. *Journal of Strategic Studies, 35*(2): 317–343.

Press-Barnathan, G. (2006). Managing the Hegemon: NATO Under Unipolarity. *Security Studies, 15*(2): 271–309.

Rapport, A. (2015). Military Power and Political Objectives in Armed Interventions. *Journal of Peace Research, 52*(2): 201–214.

Reveron, D. S. (2002). Coalition Warfare: The Commander's Role. *Defense & Security Analysis, 18*(2): 107–121.

Richter, A., & Webb, N. J. (2014). Can Smart Defense Work? A Suggested Approach to Increasing Risk- and Burden-Sharing Within NATO. *Defense & Security Analysis, 30*(4): 346–359.

Ringsmose, J. (2010). NATO Burden-Sharing Redux: Continuity and Change After the Cold War. *Contemporary Security Policy, 31*(2), 319–338.

Rynning, S. (2012). *NATO in Afghanistan—The Liberal Disconnect*. Stanford, CA: Stanford University Press.

Sandler, T., & Shimizu, H. (2014). NATO Burden Sharing 1999–2010: An Altered Alliance. *Foreign Policy Analysis, 10*(1), 43–60.

Schneider, G., & Weitsman, P. (1997). Eliciting Collaboration from "Risky" States: The Limits of Conventional Multilateralism in Security Affairs. *Global Society, 11*(1), 93–110.

Snyder, G. H. (1984). The Security Dilemma in Alliance Politics. *World Politics, 36*(4), 461–495.

Snyder, G. H. (1997). *Alliance Politics*. Ithaca, NY: Cornell University Press.

Taliaferro, J. W. (2001/2002). Security-Seeking Under Anarchy: Defensive Realism Reconsidered. *International Security, 25*(3): 152–186.

Walt, S. (1987). *The Origins of Alliances*. Ithaca, NY: Cornell University Press.

Waltz, K. (1979). *Theory of International Politics*. Boston: McGraw-Hill.

Weitsman, P. A. (1997). Intimate Enemies: The Politics of Peacetime Alliance. *Security Studies, 7*(1), 156–193.

Weitsman, P. A. (2003). Alliance Cohesion and Coalition Warfare: The Central Powers and Triple Entente. *Security Studies, 12*(3), 79–113.

Weitsman, P. A. (2004). *Dangerous Alliances—Proponents of Peace, Weapons of War*. Stanford, CA: Stanford University Press.

Weitsman, P. A. (2010). Wartime Alliances Versus Coalition Warfare. *Strategic Studies Quarterly, 4*(2), 113–136.

Weitsman, P. A. (2014). *Waging War: Alliances, Coalitions, and Institutions of Interstate Violence*. Stanford, CA: Stanford University Press.

CHAPTER 9

Domestic and Governmental Politics

In the previous chapter, we theorized how the outside-in approach of alliance politics might help explain coalition members' inclination to apply national reservations on the use of force on their coalition contributions. In the present chapter, focus shifts to the inside-out approach of governmental and domestic politics. We elaborate on multiple theoretical arguments as to how caveats may facilitate the construction of domestic winning political coalitions required to participate in coalition operations (Frost-Nielsen 2016: 29–33; 2017: 4–6). How may attributes of government, governmental institutions, and domestic politics explain the adoption of caveats in foreign policy and patterns of caveats in coalition forces?

In democratic politics, the decision-making group and networks of government ministries and agencies responsible for making foreign policy decisions of relevance to the participation in coalition forces need to win political support in government, parliament, and the public at large for policy decisions. If for no other reason, to reduce the political risk and costs if the decision to contribute forces should somehow backfire (e.g., collateral damage, own casualties) and come back to haunt the government in the future.

We have referred to Robert D. Putnam's scholarship in previous chapters. Putnam's two-level game modeling of the interaction between international and domestic politics is interesting for the purpose of the study of foreign policy and caveats because it invites us to look closer at how national reservations on the use of force may reconcile demands and expectations from alliance partners with what is possible to win support

© The Author(s) 2019   145
G. Fermann, *Coping with Caveats in Coalition Warfare*,
https://doi.org/10.1007/978-3-319-92519-6_9

for and have ratified in domestic political institutions (Putnam 1988). Putnam shares with the approach of Foreign Policy Analysis (FPA) the notion that foreign policy-making is pitched between a rock and a hard place in this sense.

What policy on caveats decision-makers come up with after assessing the international ramifications (the particular threat the coalition is to counter; the security dilemma in alliance politics) is also influenced by what the decision-makers simultaneously have assessed it might be possible to gain domestic political support for. In principle, Putnam's model and FPA allow for the symmetrical interpretation: How decision-makers understand the situation on one negotiating table may influence how the decision-makers perceive the facts and scope for political maneuvering (SPM) on the other negotiating arena.

Juliet Kaarbo notes that the analytical distinction made between external and domestic influences on states' foreign policy-making processes may be more analytical than real because of the back-and-forth reflexive process of two-level negotiations (Kaarbo 2015: 207). Still, to assess the complexities of foreign policy-making, it is necessary to reason and identify what aspects of foreign policies explained on what level of analysis, and through what causal mechanisms and paths (Zakaria 1992: 198).

Andrew Moravcsik's scholarship is a particularly useful point of departure to understand how foreign policy-making conditioned by the nature of societal forces and the aggregation of such forces in domestic politics. The fundamental assumption is that interest-crystalizing sentiments and inclinations originate and come into being prior to politics, in society. According to Moravcsik, a state's behavior toward the global arena results from the interaction between politically organized actors that represent social groups within society, and which operates within or toward the political institutions that regulate how interests are aggregated in society (Moravcsik 1997). This line of reasoning very much corresponds to the notion that "all politics is local" as discussed in Chapter 7 (O'Neill and Novak 1987).

To better understand how national reservations on the use of force in coalition operations (caveats) can balance conflicting domestic considerations, Moravcsik's inside-out approach to foreign policy-making invites us to (i) identify what domestic interests are represented in the political institutions and the foreign policy-making process; (ii) how institutions represent opportunities and limitations for different actors' ability to pursue their interests; and (iii) to study how this impact the foreign policy-making process and outcomes.

Starting from the assumption that caveats are the result of necessary compromises in domestic politics in support of participation in coalition operations (Auerswald and Saideman 2014), the present chapter offers a review of literature that study how domestic politics impacts foreign policy-making processes, as well as discussing a handful of studies researching coalition participation in particular. The theoretical discussion leads to the deduction of 23 probing hypotheses on domestic and governmental factors that may condition the foreign policy-making state's use of caveats in coalition forces.

RECONCILING DISAGREEMENTS IN DOMESTIC POLITICS

After the end of the Cold War, the scholarly interest as to how domestic politics and governmental institutions influence foreign policy-making has soared (Kaarbo 2015: 189; Schultz 2012: 478). There are studies on how political institutions impact states' willingness to participate in coalition forces. Some studies indicate that a high degree of autonomy of decision-makers in relation to other political institutions increases the likelihood that the state will contribute substantial military forces to the coalition (Auerswald 1999, 2004; Baltrusaitis 2010). Other strains of research find that external political pressure is a deciding factor in states' decision to participate in coalition operations, while domestic conditions decide *how* states contribute (Bennett et al. 1997; Mello 2014; Resnick 2013; Schuster and Maier 2006). In a recent comparative study of the application of caveats on Denmark's, the Netherland's, and Norway's contributions to NATO's intervention in Libya in 2011, Per Marius Frost-Nielsen concludes that "domestic factors help to explain whether or not there will be caveats, while external pressures help to explain the form that such caveats take" (2017: 22).

These somewhat equivocal findings nevertheless indicate that the application of caveats not only may contribute to balance the dual fears of entrapment and abandonment in alliance politics, but also contribute to the reconciliation of domestic political disagreements using a policy instrument that retracts from the initial decision to allocate substantial military forces to coalition operations.

Moreover, the literature makes useful distinctions between diplomatic, economic, and military support to coalitions (Baltrusaitis 2010; Bennett et al. 1997; Davidson 2011). It is a limitation, however, that neither of the studies distinguish between different kinds of military support that would have made it possible to grasp the nuances within this category. It is a typological limitation precisely because caveats is an expression of variation in military support to the coalition. In the lack of a fine-grained category of military support, it is difficult to read anything precise out of the literature that illuminates how national reservations on the use of force may contribute to the domestic ratification of participation in coalition forces.

Frost-Nielsen observes that it is full agreement in research on the influence of domestic and institutional factors on decisions related to the use of military force that the more actors involved in the decision-making, the more difficult it is to reach and implement an agreement (2016: 31). This finding is particularly robust for democratic regimes where decision-makers' freedom of action is contingent on approval from a political majority (Maoz and Russett 1993; Reiter and Stam 2002). David P. Auerswald and Stephen M. Saideman's research confirm this finding:

> To get an agreement past all of the potential veto players (usually the representatives of the parties in the coalition), restrictive delegation [caveats] may be the only way for the nation to participate in the military intervention. [T]he broader the ideological spectrum represented in the coalition, or the larger the number of coalition parties, the more compromises are required. (Auerswald and Saideman 2014: 77)

Auerswald and Saideman find strong empirical support for the notion that national reservations on the use of force is a policy instrument that can be used to bring about a compromise solution that allows for military participation. They argue that caveats have this reconciliatory function potentially in all forms of constitutional government. However, national reservations on the use of force are more likely to facilitate compromises where they are most needed and hardest to arrive at—in coalition governments in multi-party parliamentary systems.

The political engineering instrument of caveats may also contribute to facilitate compromises under other decision-making circumstances. Bureaucratic actors with competence and decision-making authority on issues related to security and defense may find the instrument of caveats useful. As may autocratic regimes where decision-makers require at least some tacit approval from crucial constituencies to cover their backs for the unexpected (De Mesquita 2006: 422–425; Hagan et al. 2001).

The need to arrive at some domestic political compromise on the issue of coalition participation may also arise between key cabinet members of the government, or between key decision-makers and their constituencies in parliament. In a string of studies, Juliet Kaarbo has researched foreign policy decision-making in coalition governments in parliamentary democracies.[1] Kaarbo (2008) assesses the research findings in this particular field as indeterminate. In an extensive study of negotiations between the members of coalition governments, she shows (2012) how complex such decision-making processes are and how difficult it is to generalize on the mechanisms involved in the construction of politically feasible solutions in foreign policy decision-making. Still, the literature gives some clues to what may condition the foreign policy-making process and outcomes—also of relevance for our discussion of caveats (Kaarbo 2012: 236–241).

INFLUENCES ON THE FOREIGN POLICY-MAKING PROCESS AND OUTCOMES

First, a party split by disagreement on war-participation is incapable of blocking or conclusively influence its partners in government. In such a political constellation, there is less reason to expect that a compromise is required or even possible, and thus less reason to expect caveats to be relevant.

Second, Kaarbo finds evidence that the nature of the issue at stake in foreign policy decision-making impacts the ability of political parties' with different views to cooperate:

> When positions were irreconcilable [...] fragmented action resulted. When the issue was indivisible [...], one side prevailed. Divisible issues [...] allowed the parties to make meaningful compromises. (Kaarbo 2012: 237)

The implication of this finding for the inclination to apply caveats in foreign policy-making is equivocal: On the one hand, we may expect that national reservations on the use of force applied when the issue is

---

[1] See Beasley and Kaarbo (2014), Cantir and Kaarbo (2012), Kesgin and Kaarbo (2010), Kaarbo (2001, 2003, 2008, 2012, 2015), Kaarbo and Beasley (2008), and Kaarbo and Cantir (2013).

divisible and open for compromise. On the other hand, the foreign policy instrument of caveats may also used to overcome the negotiating deadlocks resulting from indivisible policy issues and irreconcilable positions. By infusing caveats into the decision-making discussion, the skilled mediator may convince the skeptical party (or parties) that applying certain reservations on how the contingent can be used in the field, implies that the issue be transformed from an indivisible one based on (say) principled resistance to war-participation to a divisible question about how to participate.

Along with convincing appeals to the humanitarian situation and justification found in International Law (Rathbun 2004), the policy instrument of caveats may be used by political entrepreneurs to reduce the political costs of key entrenched parties, and thus show them a feasible way to make a political compromise. The significance of framing and the construction of new narratives should not be underestimated in the process of transforming the structure of negotiations from one of zero-sum to one of variable-sum (Festinger 1957; Tversky and Kahneman 1974). As is the significance of facilitators capable of constructing a larger SPM using narratives demonstrating the possible game-changing implications of caveats and other policy instruments. Arguably, the existence of mediators capable of introducing caveats into the discussion on coalition participation may have rescued governments squeezed between the rock and a hard place of domestic and international politics.

Third, institutions matter. Kaarbo (2012: 238) finds that it does matter whether foreign policy decisions are made in this or that ministry, or in institutions such as the UN or NATO for how disagreement within the government on coalition participation are solved. As to the particular foreign policy instrument of caveats, if and how institutional environments matter is open to empirical scrutiny. Still, an argument can be made that a ministry of foreign affairs is more inclined to seek a compromise using caveats, then is a ministry of defense—everything else being equal. The justification for such a proposition is, first, that ministries of foreign affairs have broader responsibility for its country's reputation around the world, and expected to go at considerable lengths applying policy instruments capable of balancing competing concerns. Second, we would expect that the military under the authority of the ministry of defense argues against caveats that require the contingent to fight with the proverbial arm tied on the back. Moreover, we may suspect that a foreign policy-making process that is close to or heavily integrated into the multilateral politics of international organizations is more inclined to produce decisions favoring unreserved coalition participation, or the minimal use of caveats.

Fourth, key decision-makers are unlikely to press matters in foreign policy if this runs the risk of overturning the government. In most cases, safeguarding the continuing existence of the government override the desire to win through with foreign policies. How may this finding relate to the application of caveats in coalition operations? Usually, national reservations on the use of force is a moderating policy instrument (restrictive caveats). This probably implies that where participation in coalition force threatens the cohesion and future of a coalition government, foreign policy decision-makers will back down from sending an unreserved military contingent. To the extent failing on the commitment to participate in the coalition is not a feasible option (recall the security dilemma in alliance politics), the government is likely to offer some reserved military contribution.

Fifth, for smaller political parties' to influence foreign policy, Kaarbo argues that they need to be consistent in how they argue their case. This finding is supported by other studies, one in which the authors also conclude that small parties need to control the ministry of foreign affairs to influence foreign policies in substantial ways (Opperman and Brummer 2014).[2] The two observations are not mutually excluding. Rather, the control over the ministry of foreign affairs is likely to facilitate some continuity and cohesion in how the party in question narrates and justify its foreign policies on behalf of the coalition government. However, it is not obvious what the implications of these findings are for the question of caveats in foreign policy-making.

For instance, it is hard to see how the size of a party, as such, influences its inclination to support the application of caveats on coalition forces. In a globalized era where the Left-Right axis in politics has

---

[2]Germany, for instance, has a long history of leaving the political responsibility of the ministry of foreign affairs ("Auswärtiges Amt") to the junior partner in coalition governments. The longest-serving minister of foreign affairs in Germany is Hans-Dietrich Genscher (1974–1992), representing The Free Liberal Democratic Party (FDP) in several governments dominated by The Christian Democratic Party CDU. Other German ministers of foreign affairs from junior coalition parties are Klaus Kinkel (1992–1998) from FDP, Joschka Fischer (1998–2005) from The Greens, and Guido Westerwelle (2009–2013) from FDP.

diminished in relevance, it is also difficult to argue that Leftist (small) parties should be more caveats-prone than (small) parties on the Right. What may make a difference for the pattern of caveats behavior, however, is the outlook and voter-basis of the party (regardless of size and ideological affiliation) on the increasingly significant Cosmopolitan–Communitarian axis of politics. While the foreign policy scope condition literature is preoccupied with the ideological Left-Right cleavage in the study of politics, the literature has little to say about how other conflict-dimensions in domestic politics may influence governments' coalition-behavior and caveats policies. Several decades ago, Political Sociologist Stein Rokkan built his theory on political development and party-formations in Europe on the assumption that there were more than the cleavage of the Left-Right axis of politics involved in the shaping of nation-states and thus their policies. Political divides based on differences in language, culture, and religion did matter for political development, as does the structuring of the relationship between the urban and the rural (Rokkan and Eisenstadt 1973; Berntzen and Selle 1990).

As mentioned, a more recent dimension in politics that transgress and cut across the traditional Left-Right ideological axis is the cleavage between parties representing a cosmopolitan versus a communitarian outlook. The cleavage is a result of the inherent economic and political tensions in globalization, especially since the end of the Cold War. Arguably, we should pay more attention to the study of whether and how "extrovert" (cosmopolitan/globalist) and "introvert" (communitarian, nationalist) governments make different decisions on coalition participation and the application of national reservations on the use of force in coalitions. The argument can be made that governments with a globalist outlook tend to assign a higher value to international cooperation in itself. Arguably, this collaborative instinct would, in turn, make globalist governments more inclined to deliver on its alliance commitments even if the security dilemma points otherwise, and thus to participate in coalition forces with few if any caveats (see hypotheses DGP-H12–15).

From this line of reasoning, we may expect parties in government with a globalist (cosmopolitan) outlook to be more intent on coalition involvement. Globalist parties would likely favor full participation in coalition forces without any strings attached (no caveats), everything else being equal. Their subsidiary position will likely be to apply caveats if the political alternative to caveats is no participation at all. As to the party with a predominantly nationalist (communitarian) outlook, we may speculate that its primary position is no participation in the coalition force, everything else being equal. The subsidiary position would likely be some degree of military participation granted that the coalition partners accepted considerable reservations on the use of force. However, we should bear in mind that regardless of outlook on the cosmopolitan-communitarian axis, any political party need to take into account the dynamics of alliance politics as discussed in the previous chapter, as well as the realities of war discussed in the subsequent chapter.

If the small party in government with a nationalist-communitarian outlook is heading the ministry of foreign affairs, we expect that the primary position of the government would be to favor a smaller military contribution to the coalition as well as to attach a degree of restrictive caveats to it. It would be a rarity though that a Center-Left and even a Center-Right government pick the minister of foreign affairs from a minor party with a distinct nationalist outlook in an era still dominated by a globalist Western outlook.

The Prerogative to Make Foreign Policies, Ideology, Decision-Making Mechanisms

In the constitutions of many states, the prerogative of making decisions on foreign policy and external security issues belongs exclusively to the government. This authority includes making decisions related to coalition participation. However, there are other considerations than the formal institutional arrangements of the state that decide what autonomy the executive branch has in foreign policy-making in relation to other state institutions (Dieterich et al. 2010; Peters and Wagner 2014). The need to develop political compromises arise between the government and the parliament in particular.

In a comparative study of democratic states' participation in the military interventions in Kosovo (1999), Afghanistan (2001–2002) and Iraq (2003–2010), Patrick A. Mello (2014) assesses variations in participation. One of the several indicators used to measure the variation in participation is national reservations on the use of force. Mello finds no empirical evidence to support the notion that parliamentary veto powers on the issue of deploying military forces abroad make a difference as to whether and how states participate in coalition forces. This finding is an indication that whether domestic negotiations over coalition participation result in caveats depends on more contextual factors.

One such contextual factor is ideology. Brian Rathbun (2004) finds that European political parties' attitude toward participating in military interventions vary with ideological affiliation. Parties with different ideological outlooks tend to interpret the "national interest" differently, and thus compete on exerting influence on the definition of the national interest and the narratives supporting the conception. According to Rathbun, parties to the Right are inclined to see military intervention as a legitimate policy instrument if the use of force serves the economic and security interests of the state in a rather direct sense. Left-leaning parties, on the other hand, tend to support the use of force in foreign policy on the condition that force executed within a multilateral framework, finds justification in International Law, and is a response to a humanitarian call for action.

How may Brian Rathbun's findings and line of reasoning inform the researcher on the application of the instrument of caveats in foreign policy? One reasonable expectation is that a Center-Left coalition government will turn down an invitation to contribute to a narrowly composed "coalition-of-the-willing" coalition, which rests on a weak foundation in International Law, and fails to produce a credible humanitarian justification. A Center-Right coalition government, on the other hand, would likely be inclined to make an unreserved military contribution if considerable economic and security interests were at stake. For a Center-Left government to accept participation in a coalition on mainly economic and security terms, significant concessions would likely have to be made in the currency of national restrictions on the use of force. Another possible compromise for a Center-Left government on the terms mentioned would be to reduce the military contribution to a symbolic one, depending on what other factors need to be taken into consideration.

Per Marius Frost-Nielsen (2016: 33) notes that Auerswald and Saideman (2014) have offered the most explicit theoretical argument for national reservations on the use of force in coalitions. They argue that caveats are the outcome of necessary political compromises in support of the participation in coalition operations. The more actors that exhibit veto powers in the negotiations, the larger is the need for a compromise solution. Auerswald and Saideman also argue that the larger the ideological differences between the actors with access to the negotiations, the more demanding it is to work out a compromise solution.

Putting emphasize on the number of veto actors involved in the negotiations and the degree of ideological compatibility between the actors is a useful point of departure for research. However, the literature referred to above indicates that the issue as to why and under what conditions states apply caveats in foreign policy-making is more complex. Research on caveats is thus likely to benefit from a broader theoretical approach and ditto research methodologies to grasp the complex domestic decision-making problems that the foreign policy instrument of caveats may contribute to solving.

By actually tracing foreign policy decision-making processes empirically, Kaarbo (e.g., 2008) finds that domestic political factors such as institutional possibilities and limitations, political struggle as to what preferences come out on top within and between parties, and the relationship between the executive and legislative powers of the state do influence foreign policy-making outputs. Frost-Nielsen (2016, 2017) has exploited such insights in empirical research on caveats within the multi-level approach of FPA. Moreover, domestic and institutional framework conditions may influence foreign policy outcomes through policy-making mechanisms such as group polarization, persuasion, and strategic influence, and psychological factors such as willingness to accept the risk. Context-specific operationalization of such factors is necessary to assess precisely how caveats come to facilitate the construction of a winning political coalition for the participation in military coalition operations.

Theorizing caveats at the domestic level of politics is crucial also within the approach of FPA although foreign policy analysts are the first to point out that in *foreign* politics not "*all* politics is local." This is especially the case for junior members of security alliances, who are inclined to adapt to international circumstances rather than posturing a pro-active foreign policy of their own. Domestic impulses interact with global politics in the considerations of foreign policy decision-makers. How, to what extent, and with what bias are empirical questions.

What can we learn about the politics of caveats from the domestic and governmental politics-relevant literature on foreign policy-making discussed above? What may the inside-out perspective on foreign policy-making add to the outside-in perspective of alliance politics discussed in the previous chapter? How may the domestic politics of foreign policy-making shed light on states' inclination to participate in coalition forces with or without particular national reservations on the use of force? How do domestic politics influence the pattern of caveats in coalition operations? These are relevant research questions at the domestic and governmental levels of analysis.

POINTING THE ARGUMENTS: EMPIRICAL PROPOSITIONS

As is appropriate in a rather nascent field of research, the answers to such questions do not come dressed as generic conclusions. Rather, what follows is the formulation of several empirical propositions as to what makes states inclined to participate in coalition forces and apply caveats. The multitude of empirical propositions reflect the several lines of reasoning in literature and is offered as points of departure for future research on the politics of caveats. At the foundation of most of the hypotheses that follow is Auerswald and Saideman's assumption that national reservations on the use of force are the outcome of necessary political compromises in support of the participation in coalition operations (2014). Somehow, depending on conditions, national reservations on the use of force is an instrument that may facilitate the construction of domestic winning political coalitions required to participate in allied coalition operations.

"Domestic and Governmental Politics" hypotheses (DGP-H1–23): Everything else being equal, we may consider that;

DGP-H1   *"If and how" hypothesis:* When assessing invitations to participate in coalition forces, states are inclined to let external considerations related to threats to security and alliance politics decide *whether* to contribute, and domestic concerns related to political feasibility decide *how* to contribute.

DGP-H2   *"Dual-purpose" hypothesis:* States apply caveats not only to balance security dilemmas in alliance politics but even more to reconcile domestic political disagreements.

DGP-H3   *"Latitude" hypothesis:* Decision-makers with a high degree of autonomy in relation to other foreign policy decision-making institutions, and with considerable discretion to make decisions on coalition participation are inclined to contribute substantial military forces to the coalition.

DGP-H4   *"Veto, no impact" hypothesis:* Parliamentary veto powers on decisions related to use of force abroad does not make a difference as to *whether* and *how* states participate in coalition forces.

DGP-H5   *"Reservations in numbers" hypothesis:* The more actors (veto-players) having access to the domestic decision-making process, the more demanding it is to reach a compromise solution on coalition participation and thus a greater inclination to apply caveats on the contribution as part of a political compromise.

DGP-H6   *"Deep compromise" hypothesis:* In democratic coalition governments in multi-party parliamentary systems, the challenge of having military coalition participation approved by the other parties in the government coalition is amplified and the inclination to apply restrictive caveats even stronger to protect the contribution to the coalition forces.

DGP-H7   *"Ideological polarization" hypothesis:* The larger the ideological distance between the veto-players in the decision-making process, the more demanding it will become to work out a compromise solution. The prospect of deadlock increases the probability that caveats be attached to force contribution.

DGP-H8   *"Legitimate cause" hypothesis I:* A Center-Left coalition government is inclined to turn down an invitation to contribute militarily to a narrowly composed "coalition-of-the-willing" coalition, which rests on a weak foundation in International Law, and fails to find support in a credible humanitarian narrative.

DGP-H9   *"Legitimate cause" hypothesis II:* A Center-Right coalition government is inclined to make an unreserved military contribution to a coalition force if considerable economic and security interests are at stake.

DGP-H10   *"Legitimate cause" hypothesis III:* For a Center-Left government to accept military participation in a coalition on mainly economic and security grounds, considerable concessions are likely to be made in the form of restrictive reservations on the use of force.

DGP-H11   *"Legitimate cause"* hypothesis *IV*: For a Center-Left government to accept military participation in a coalition on mainly economic and security grounds, the military contribution would likely be reduced to a token one, depending on what other concerns need to be taken into consideration.

DGP-H12   *"Cosmopolitan outlook"* hypothesis *I*: The inclination and primary position of a coalition government party with a cosmopolitan/globalist outlook is to accept full participation in coalition force without any strings attached (no caveats).

DGP-H13   *"Cosmopolitan outlook"* hypothesis *II*: The subsidiary position of a coalition government party with a cosmopolitan/globalist outlook would be to prefer the application of caveats on the military contribution to no participation at all.

DGP-H14   *"Communitarian outlook"* hypothesis *I*: The inclination and primary position of a coalition government party with a predominantly communitarian/nationalist outlook is to decline any participation in coalition force.

DGP-H15   *"Communitarian outlook"* hypothesis *II*: The subsidiary position of a coalition government party with a communitarian/nationalist outlook is to accept some degree of military participation provided the government coalition partners accept considerable restrictive caveats.

DGP-H16   *"No unity, no leverage"* hypothesis: A party split by disagreement on war participation and thus incapable of blocking or conclusively influence its governmental partners, is unlikely to gain acceptance for the application of caveats.

DGP-H17   *"Unlikely case"* hypothesis: A small party in a coalition government, with a communitarian/nationalist outlook on foreign affairs and in control of the ministry of foreign affairs, is inclined to favor a smaller military contribution to the coalition and to attach a degree of restrictive caveats to it.

DGP-H18   *"Institutional prerogative"* hypothesis: The inclination to apply caveats on force contributions is stronger if the decision-making power rests in the ministry of foreign affairs than in the ministry of defense.

DGP-H19   *"Institutional connect" hypothesis*: The inclination to apply caveats on force contributions is weaker if the foreign policy-making process is close to or strongly integrated into the multilateral politics of security organizations.

DGP-H20   *"Variable-sum" hypothesis*: The more divisible and structurally open for compromise the issue of coalition participation is, the more inclined is the coalition government to apply caveats on the military contribution.

DGP-H21   *"Zero-sum" hypothesis*: The more indivisible and structurally closed for compromise the issue of coalition participation is, the less inclined is the coalition government to agree upon a military contribution.

DGP-H22   *"Game changer" hypothesis*: The more skilled the political entrepreneur (mediator) is in framing the political instrument of caveats as a tool capable of transforming the contentious issue of participating in coalition force from one of zero-sum (indivisible) to one of variable-sum (divisible), the more likely it is that restrictive caveats are applied to secure coalition participation.

DGP-H23   *"Facing extinction" hypothesis*: Where the future of the government is at stake on the question of coalition participation, foreign policy decision-makers will either back down from participating in the coalition operation or—depending on circumstances—offer a minimal force or a force with considerable restrictions on the use of force.

This finalizes the formulation of empirical implications of the several lines of reasoning discussed in this chapter. We will not make any attempt to operationalize the host of independent variables included in the several hypotheses.

Almost all hypotheses are related to attributes of the political institutions involved (governments, ministries, parties), issues at stake in making decisions on the military participation in coalition forces, and how these issues resonate with domestic politics. However, the literature has less to say on the likely influence of media, non-governmental organizations and public opinion on the SPM decision-makers have at their disposal when making decisions on the extent and quality of coalition participation (Kreps 2010).

An exception is again Auerswald and Saideman's study of national reservations on the use of force in coalition operations in Afghanistan. Based on their data, no causal inference be established between public opinion support for coalition participation and the pattern of caveats applied (2014: 19–21). This lack of connection may be because information on coalition participation is more widely distributed to the public through media channels, than any information on caveats, which tends to go under the radar of the media, if at all released to the public by the authorities.

Regardless of preliminary findings and theoretically informed speculation, much remains to be researched on how society through media, non-governmental organizations and networks, and the public at large impact politicians' decision on coalition participation. In this lies an invitation to look into what the extensive fields of Political Behavior and Political Communication may contribute to the study of coalition behavior.

References

Auerswald, D. P. (1999). Inward Bound: Domestic Institutions and Military Conflicts. *International Organization, 53*(3), 469–504.

Auerswald, D. P. (2004). Explaining Wars of Choice: An Integrated Decision Model of NATO Policy in Kosovo. *International Studies Quarterly, 48*(3), 631–662.

Auerswald, D. P., & Saideman, S. M. (2014). *NATO in Afghanistan: Fighting Together, Fighting Alone.* Princeton, NJ: Princeton University Press.

Baltrusaitis, D. F. (2010). *Coalition Politics and the Iraq War.* Boulder, CO: First Forum Press.

Beasley, R. K., & Kaarbo, J. (2014). Explaining Extremity in the Foreign Policies of Parliamentary Democracies. *International Studies Quarterly, 58*(4), 729–740.

Bennett, A., Lepgold, J., & Unger, D. (Eds.). (1997). *Friends in Need—Burden Sharing in the Persian Gulf.* New York, NY: St. Martin's Press.

Berntzen, E., & Selle, P. (1990). Structure and Social Action in Stein Rokkan's Work. *Journal of Theoretical Politics, 2*(2), 131–149.

Cantir, C., & Kaarbo, J. (2012). Contested Roles and Domestic Politics: Reflections on Role Theory in Foreign Policy Analysis and IR Theory. *Foreign Policy Analysis, 8*(1), 5–24.

Davidson, J. W. (2011). *America's Allies and War: Kosovo, Afghanistan, and Iraq*. New York, NY: Palgrave Macmillan.

De Mesquita, B. B. (2006). *Principles of International Politics: People's Powers, Preferences, and Perceptions*. Washington, DC: Congressional Quarterly.

Dieterich, S., Hummel, H., & Marschall, S. (2010). *Parliamentary War Powers: A Survey of 25 European Parliaments*. Geneva: Geneva Centre for the Democratic Control of Armed Forces (DCAF).

Festinger, L. (1957). *A Theory of Cognitive Dissonance*. Stanford, CA: Stanford University.

Frost-Nielsen, P. M. (2016). *Betingede forpliktelser. Nasjonale reservasjoner i militære koalisjonsoperasjoner*. Ph.D. Dissertation in Political Science, Department of Sociology and Political Science, Norwegian University of Science and Technology (NTNU), Trondheim.

Frost-Nielsen, P. M. (2017). Conditional Commitments: Why States Use Caveats to Reserve Their Efforts in Military Coalition Operations. *Contemporary Security Policy, 38*(3), 371–397.

Hagan, J. D., Everts, P. P., Fukui, H., & Stempel, J. D. (2001). Foreign Policy by Coalition: Deadlock, Compromise, and Anarchy. *International Studies Review, 3*(2), 169–216.

Kaarbo, J. (2001). Domestic Politics and International Negotiations: The Effects of State Structures and Policy-Making Processes. *International Interactions, 27*(1), 169–205.

Kaarbo, J. (2003). Foreign Policy Analysis in the Twenty-First Century: Back to Comparison, Forward to Identity and Ideas. *International Studies Review, 5*(1), 156–163.

Kaarbo, J. (2008). Coalition Cabinet Decision Making: Institutional and Psychological Factors. *International Studies Review, 10*(1), 57–86.

Kaarbo, J. (2012). *Coalition Politics and Cabinet Decision Making*. Ann Arbor, MI: University of Michigan Press.

Kaarbo, J. (2015). A Foreign Policy Analysis Perspective on Domestic Politics Turn in IR Theory. *International Studies Review, 17*(2), 189–216.

Kaarbo, J., & Beasley, R. K. (2008). Taking It to the Extreme: The Effect of Coalition Cabinets on Foreign Policy. *Foreign Policy Analysis, 4*(1), 46–81.

Kaarbo, J., & Cantir, C. (2013). Role Conflict in Recent Wars: Danish and Dutch Debates over Iraq and Afghanistan. *Cooperation and Conflict, 48*(4), 465–483.

Kesgin, B., & Kaarbo, J. (2010). When and How Parliaments Influence Foreign Policy: The Case of Turkey's Iraq Decision. *International Studies Perspective, 11*(1), 19–36.

Kreps, S. E. (2010). Elite Consensus as a Determinant of Alliance Cohesion: Why Public Opinion Hardly Matters for NATO-led Operations in Afghanistan. *Foreign Policy Analysis, 6*(3), 191–215.

Maoz, Z., & Russett, B. (1993). Normative and Structural Causes of Democratic Peace, 1946–1986. *American Political Science Review, 87*(3), 624–638.

Mello, P. A. (2014). *Democratic Participation in Armed Conflict: Military Involvement in Kosovo, Afghanistan, and Iraq.* New York, NY: Palgrave Macmillan.

Moravcsik, A. (1997). Taking Preferences Seriously: A Liberal Theory of International Politics. *International Organization, 51*(4), 513–553.

O'Neill, T., & Novak, W. (1987). *Man of the House: The Life and Political Memoirs of Speaker Tip O'Neill.* London: Random House.

Opperman, K., & Brummer, K. (2014). Patterns of Junior Partner Influence on the Foreign Policy Coalition Governments. *British Journal of Politics and International Relations, 16*(4), 555–571.

Peters, D., & Wagner, W. (2014). Executive Privilege or Parliamentary Proviso? Explaining the Sources of Parliamentary War Powers. *Armed Forces and Society, 40*(2), 310–331.

Putnam, R. D. (1988). Diplomacy and Domestic Politics: The Logic of Two-Level Games. *International Organization, 42*(3), 427–460.

Rathbun, B. C. (2004). *Partisan Interventions—European Party Politics and Peace Enforcement in the Balkans.* Ithaca, NY: Cornell University Press.

Reiter, D., & Stam, A. C. (2002). *Democracies at War.* Princeton, NJ: Princeton University Press.

Resnick, E. N. (2013). Hang Together or Hang Separately? Evaluating Rival Theories of Wartime Alliance Cohesion. *Security Studies, 22*(4), 672–706.

Rokkan, S., & Eisenstadt, S. N. (1973). *Building States and Nations.* London: Sage.

Schultz, K. (2012). Domestic Politics and International Relations. In W. E. Carlsnaes, T. E. Risse, & B. A. Simmons (Eds.), *Handbook of International Relations* (pp. 478–502). London: Sage.

Schuster, J., & Maier, H. N. (2006). The Rift: Explaining Europe's Divergent Iraq Policies in the Run-Up of the American-Led War on Iraq. *Foreign Policy Analysis, 2*(3), 223–244.

Tversky, A., & Kahneman, D. (1974). Judgment Under Uncertainty: Heuristics and Biases. *Science, 185*(4157), 1124–1131.

Zakaria, F. (1992). Realism and Domestic Politics: A Review Essay. *International Security, 17*(1), 177–198.

CHAPTER 10

Politics of Implementation

In the two previous chapters, we discussed the foreign policy instrument of national reservations on the use of force in terms of how caveats may solve decision-making problems in alliance politics and domestic politics. Common to outside-in and inside-out explanations of foreign policy are their preoccupation with the initial political decision-making problems of whether and how to contribute to coalition forces.

What we have left out so far, is reasoning on how caveats may help the troop-contributing nation to assert political control and manage the actual execution of the military mission in the theater of war (Frost-Nielsen 2017: 7–9; 2016: 34–40). First, we discuss a selection of scholarships in principal-agent theory to shed light on why political decision-makers need to invest time and resources to make sure that political intentions adequately implemented in military action in the area of deployment. This part addresses the age-old question of how to politically control the sword, and thus secure that war continues staying politics with other means.

Second, we argue that the considerable complexity of the multilateral military organization increases the potential for mission creep and shirking in the military implementation of political mandates. Political concerns about goal drifting in the operational implementation of the mandate is a strong incentive for decision-makers to put in place arrangements for the oversight of the military.

Third, we theorize on how the instrument of national reservations on the use of force is applied to facilitate political control of

© The Author(s) 2019
G. Fermann, *Coping with Caveats in Coalition Warfare*,
https://doi.org/10.1007/978-3-319-92519-6_10

military implementation. Prospects theory is discussed to show how decision-makers by framing coalition participation in terms of prospects for loss or gain vary in their inclination to accept risks, and thus in their disposition to apply caveats to military contingent in coalition operations.

Finally, we point theoretical arguments into fourteen empirical propositions on the trade-off between political and operational concerns, as well as what condition the application of caveats in the bureaucratic politics of implementation.

Recall the promise of Foreign Policy Analysis (FPA) to mobilize theory on multiple levels of analyses to explain foreign policy outcomes. Recall also the process-tracing emphasize of the FPA approach that invites the theoretically informed study of decision-makers' scope for political maneuvering (SPM), the negotiation of policy goals and instruments among veto-players at different levels of politics, the justification of policies in the eyes of diverse political constituencies, and the actual implementation of policies agreed on. The present chapter closes this dual effort in the study of the foreign politics of caveats by taking a closer look at the implementation process, in particular, the link between the domain of political decision-making and the operational level of military implementation.

THE AGENCY PROBLEM IN CIVIL-MILITARY RELATIONS

Having negotiated a decision to contribute militarily to the coalition force, the coalition member *in spe* is finally in a position to make good on his word. It is time to implement the force contribution. Even without experiencing any political resistance in the labyrinths of alliance politics and domestic politics, the state still may have reason to apply reservations on the use of force as a means of securing some measure of national political control over the military contribution. Without political controls, the military contingent may drift and become to operate outside the political parameters of the settlement. After all, the policy-makers do not themselves implement their decisions.

In all areas of public service, decision-makers are more or less dependent on the expert advice from bureaucrats and agencies with relevant competencies. The more dependent politicians are on the advice and services of implementing agencies, the more influence the facilitating and implementing bureaucracy has over decision output and implementation (Weber 2010: 129–132). Thus, a principal–agent relationship comes into being between some commander in the field (an implementing "agent") who holds an information advantage over the political or bureaucratic decision-maker (a "principal") in his or her capacity to factually inform decisions and implement policies. The principal, who possesses the formal and political authority, cannot do without the decision-preparing expertise and implementing skills of the agent who nominally is subordinated to the principal.

Two fundamental challenges arise from the principal–agent relationship as described. First, the agent's information advantage may result in the implementation of political decisions in such a way as to deviate from and even disagree with the principal's intentions. Recall from Chapter 6 that there is no necessary one-to-one relationship between political intentions (choice of aims and means) and the implementing outcome of such policies. Second, the politically responsible principal with the formal authority to check the agent, needs to find ways to motivate and induce the better informed and skilled implementing agent to behave in the same way as the decision-making principal would have if he possessed the same information and implementing skills as the agent in question (Feaver 1998: 408–409).

Different versions of the principal-agent approach have been developed for different research purposes (Miller 2005). For the present study, it is of particular relevance to review how the literature on civil–military relations are theorizing principal–agent relationships between political decision-makers and military implementers (Auerswald and Saideman 2014: 54–56; Feaver 1999; Goodpaster and Huntington 1977; Huntington 1957). As in other realms of public policy, political decision-makers in charge of security and defense policies depend on the delegation of authority to public servants to implement policies. In the context of contributions to coalition forces, the public servant and implementing agent is the military.

Per Marius Frost-Nielsen emphasizes that the implementation of military means is different from any other political instrument in that it involves the systematic use of armed force with potentially irreversible material and human destruction (2016: 34). Much can go wrong in military operations in theaters of war. Thus, the use of armed force implies an inescapable tension between two contradictory concerns and logics: On the one hand, the requirement of effective decision-making and rapid adaptation to evolving circumstances in the field. On the other, the more time-consuming requirement of maintaining political control of the operation to secure that the implementation harmonizes with the political intentions bringing about the military deployment in the first place (Greentree 2013: 326; Cohen and Gooch 2006; George 1991).

TENSIONS BETWEEN THE LOGIC OF WAR AND THE NATURE OF POLITICS

In a military unit, individual soldiers and officers put their lives at risk by committing to collective military aims at the tactical level. To overcome individual fears and hesitations that may reduce the military efficiency of collective action, the operational behavior of military units in battle is up to a point automated by drilled rules and standardized procedures (King 2006; Soeters et al. 2010: 1–6). To rely on rehearsed behavior as a response to orders given in potentially life-threatening circumstances, may reduce the militaries capacity to keep in focus how military action continues to serve the underlying political rationale for the operations.

Regardless of how military operations are executed, military action impacts the short- and long-term prospects for escalation and de-escalation and the probability for long-term political goal-attainment. For instance, military solutions with a short-term tactical gain may have the unintended consequence of producing a negative strategic effect by escalating and prolonging the military conflict (Hasik 2014; Meyer 2013; Ruffa 2013; Ruffa et al. 2013; Scheipers 2014; Sookermany 2012). Precisely because military action on the tactical level may have far-reaching political consequences, political principals have strong incentives to extend their influence beyond the planning of military operations. Policy-makers also need to monitor and control how the military agents are implementing operations (Ruffa et al. 2013).

Due to their information and skills-advantage, military agents on the other side of the principal–agent relationship have an incentive to dodge political control and ongoing direction (Feaver 2003: 58–68). Such shirking is made possible by two different information problems facing the decision-making principal: One relates to the principal's inadequate knowledge of the agent's real preferences, the so-called "adverse selection problem." The other regards the principal's imperfect access to the agent's military actions on the ground, the so-called "moral hazard problem" (Rauchhaus 2009: 874–876).

Different suggestions have surfaced as to why civilian and military authorities may have diverging preferences resulting in goal drifting, and how civilian authorities may control and cope with the dual information problem. Barry Posen (1984), in particular, argues that any military organization is inclined to prefer offensive military solutions to deny the enemy the military initiative in warfare. Political principals, on the other hand, are inclined to prefer defensive military solutions to preserve political options and avoid the unwanted escalation of the conflict.

According to Posen, the militaries' inclination toward offensive operations does not imply that military organizations favor the use of force in all circumstances, but rather that the military never can rule out warfare as a possibility and offensive operations as the preferred approach to battle. After all, it is the function of the military to prepare for military conflict if political circumstances call for such actions. For the military agents, the inclination toward offensive military action is a strategy to decrease uncertainty and improve the odds of military combat. Taking the initiative in military operations is an investment in controlling the premises for military combat and push the development of warfare in the preferred direction.

From this line of reasoning, we expect national military agents in coalition forces to favor no national restrictions on the use of force whatsoever. To the extent the military act as veto-players in the planning of and decision-making on the participation in the coalition force, this is likely to increase the inclination for the military contingent to deploy on unreserved terms. To the extent restrictive caveats are applied, we expect the implementing military agent to work for a flexible interpretation of the national reservations. These are both reasonable interpretations of what we dub the "no-arm-tied-on-the-back" hypothesis. That is the notion that military commanders at the level of the national contingent do not wish to fight a war with one arm tied to the back due to the imposition of hard-to-cope-with restrictions on the use of force.

Barry Posen's research on civil–military relations has triggered scholarly debate. Elizabeth Kier (1997) argues that the military's preference for offensive military action should not be taken for granted. Rather, the military doctrines of national military institutions depend on particular military culture. Moreover, Jack Snyder (1984) argues from a step further back in the causal chain that military and political preferences are historically conditioned. Snyder's argument finds empirical support in Tony

Ingesson's study of "one of the most trigger-happy UN units in Bosnia" (2017). Ingesson finds that the trigger-happiness of the Swedish-lead and -dominated Nordic battalion was conditioned by a "well-entrenched culture of mission command" that emphasized offensive action and delegated a considerable degree of autonomy to the military commander to make operational decisions on the spot. This culture of mission command was institutionalized in a Cold War environment where Swedish forces were expected to continue fighting against Soviet forces even if the lines of communication to headquarter were cut, and also if rumors of capitulation were substituting access to solid information. In the complicated context of Bosnia, this energetic military doctrine translated into a Nordic battalion that confronted Serb forces whenever the Swedish battalion commander considered it necessary, thus deviating from the considerably more evasive military posture practiced in other contingents in UNPROFOR (Ingesson 2016). Finally, Shawn Cochran's research (2014) indicates that military leaders' preferences as to military action (offensive or defensive) also depend on the militaries' assessment of the legitimacy of the military intervention in question.

The requirement at this point is not to establish what militaries' real preferences on offensive actions are in a particular context, but rather to confirm that political principals and military agents may have diverging preferences on the issue of the use of force. Such discrepancy implies that political intentions may be lost in the implementing translation. This is, as mentioned, precisely the reason why politicians want to exert some control over the military implementation of the mandate. In most cases, it is safe to assume that policymakers and the military agree on the overarching political purpose of the use of force. Where disagreement may persist is more likely to relate to what calibration of military force is necessary to fulfill the political objectives, and what level of risk-taking is appropriate in battle (Feaver 2003: 59).

Hence, the political requirement to control the military implementation does not necessarily rest on the suspicion that the military will want to evade political direction. More likely, political precaution relates to a concern that the military in the capacity of its particular professional competence perceives and assesses the problem-solving situation differently than the political principal, and that the military implementation of policies thus may deviate from the political rationale (Frost-Nielsen 2016: 36). This potential disagreement between political and military outlooks relates to and partly overlaps with the distinction made in Chapter 3 between "prior" (general) and "local knowledge" in the context of exercising situational judgment in the field (Osiel 1999). Arguably, the more specialized a military unit or an officer is, the more restricted is the professional framework for analyzing the problem in question, and the fewer the means available to cope with the task (adverse selection problem). Operational tunnel vision increases the probability of goal drifting and the chance that political intentions become lost in military implementation.

Adding to the complexities of the principal–agent relationship in the politics of implementation of military force in coalition operations is the reality of how military organizations function in battle. As mentioned, the "moral hazard problem" relates to the political principal's inability to directly monitor and assess the use of force in the theater of war. For instance, for the use of force to be in accordance with International Law (Jus in Bello), the military utility of attacking a target must be greater relative to the risk of civilian casualties. To assess whether this fundamental tenet of International Law is fulfilled the political principal has to rely on information and assessments from the military leaders.

In such a situation, the military agent may use his information-advantage to manipulate the narrative of events to serve own preferences. Less sinister, the military agent will report the incident according to his best professional judgment. Still, the political principal may have assessed the situation differently if they had the similar competence and full access to operational information (Cronin 2013; Kahl 2007). This is because the assessment of facts not only depends on the facts themselves but as much on the interpretive frameworks and the particular role of those actors assessing the facts. Recall from Chapter 7, "where you stand, depends on where you sit."

The challenge confronting political principals in the implementation of military operations designed to serve political purposes thus presents itself as a dilemma. Due to their dependence upon military expertise, political decision-makers must delegate considerable authority to the military to translate political intentions into military action. However, the information advantage of the implementing military agent creates uncertainty in the political principals as to whether the political intentions be best served by the implementing solutions the military advice and act upon. In coping with the decision dilemma and balancing the two diverging considerations, the political decision-makers have a strong incentive to monitor, supervise and control the military implementation by various means. One such means is caveats.

CAVEATS IN THE POLITICS OF IMPLEMENTATION: ORGANIZATION, RISK ACCEPTANCE, PROSPECTS FOR LOSS AND GAIN, THE DELEGATION OF IMPLEMENTING AUTHORITY

In coalition operations, the political challenge of controlling the sword amplifies. Where a national military unit is under command in a military coalition, the military unit often has to act on orders from officers from other nations. The national contingent may even be integrated into other nations' forces as tactical battle units (Soeters and Manigart 2008: 2–3). This implies that the national political intentions that justify the military contribution now be filtered through an extremely complex military organization—a "fuzzy set organization"—with multiple levels, and several nations' military agents and political principals on board (Coletta 2013).

In such an environment, Frost-Nielsen argues, national political authorities face an even greater challenge in monitoring and controlling that their military contingents act in accordance with political intentions (2016: 37). Differences in culture, language and military doctrines between nations' military representatives in the coalition result in a complex set of only partly overlapping preferences and interpretive frameworks (Rosen 1995; Ruffa 2013, 2014; Soeters and Manigart 2008). Everything else being equal, this heterogeneous institutional and mediating link between national political principal and military agent represents a serious challenge for national political authorities to monitor and control respective military contingents.

How then may national reservations on the use of force alleviate this state of affairs in the eyes of the national principal? In a complex multinational military context, the foreign policy instrument of caveats may introduce an element of political supervision by restricting what the national contingent allowed to do without the prior knowledge and explicit consent of the national political authorities. The UN-lead operation in Somalia in the 1990's (UNOSOM II) may serve to illustrate the usefulness of caveats in this regard. In UNOSOM II, national contingents interpreted the UN mandate differently. Several national contingents were subject to the requirement that orders given through the UN chain of command be approved by the national government before implementation (Chopra et al. 1995: 88; Von Hippel 2000: 75).

Caveats may also enhance the political capacity to control military action in the field. German units in NATO's ISAF coalition in Afghanistan long had national reservations that limited the amount of force German soldiers were allowed to use in self-defense (Noetzel and Rid 2009: 75–76). A case of permissive caveats is the Danish contingent in ISAF that was instructed by the Danish government to ignore orders from NATO and use additional force if Danish troops found it necessary to defend themselves more effectively (Knudsen and Klingenberg 2013: 30).

We have established that there is no automatic one-to-one relationship between political intentions and operational implementation. In Chapter 6, we discussed how the ideal-typical model of policy-making and implementation often break down because "the reality of policy-making is extremely messy" (Clarke 1996: 27). The risk of inconsistency between national political intentions, and what the coalition Force Commander might order the national contingent to do is always there for the political principals to worry about. In this context, how may risk perception and risk acceptance influence decision-makers inclination to apply caveats to the force contribution?

Very much in line with FPA reasoning, Joseph Lepgold and Brent Sterling (2000) distinguish between decision-making risks at two different levels of analyses in relation to use of armed force. One relates to international repercussions resulting from any failure to achieve the political objectives justifying the war effort. The other concern is the risk that the war effort threatens parliamentary support for the government. The more willing policymakers are to accept the potential repercussions from domestic politics, the less inclined they are to apply restrictive caveats. Conversely, the more willing national decision-makers are to accept the risk of being singled out for criticism by allies, the more inclined they are to offer a limited military contribution or apply restrictive caveats.

However, this line of reasoning does not explain why some decision-makers are more risk prone or risk averse than others. The literature on policy-making and crisis management suggests that risk acceptance is influenced by several factors, including support from public opinion and political constituencies (Boettcher III 2004); how decision-makers assess the effectiveness of use of force in relation to the political objectives (Lepgold and Sterling 2000); and by certain psychological attributes of key decision-makers (Kowert and Hermann 1997). Of particular interest is Jack S. Levy's prospects theory, which relates risk acceptance to how decision-makers are inclined to frame the prospects for loss and gain (Levy 1996, 1997).

Based on insights from psychological research, prospects theory posits that most policy-makers are risk averse when facing an opportunity for gain as compared to status quo, but more risk accepting when confronted with prospects of loss. In defending status quo, decision-makers are more risk accepting, but more risk averse when trying to offset status quo to own advantage. Hence, the offensive instinct is more cautious while the defensive instinct of securing own turf invites more risk-accepting behavior. This insight strikes an intuitive note.

Applied to the context of military coalitions, we might expect decision-makers who are inclined to frame the decision to participate in prospects for loss be more risk-acceptant and thus inclined to abstain from tight controls on the actions of national contingents. On the other hand, decision-makers who frame their decisions in terms of an opportunity for gain will be more risk averse and careful, and thus use caveats as a means of establishing back channels for monitoring and directly controlling the actions of the contingents. How, then, may risk averse, troop-contributing policy-makers, knowing that they are pushing the envelope of status quo, apply caveats to decrease their exposure to risk and uncertainty in the phase of military implementation?

Recall that the division of labor between political principals and military agents makes it difficult for the principal to keep up to date on the military agents' true preferences and conduct in battle. This kind of uncertainty creates a strong incentive for political authorities to monitor military implementation. The information gap between principal and agent is real for any military organization but amplified in complex coalition forces. With a risk averse mindset, decision-makers have an even stronger incentive to monitor and apply national reservations on the use of force.

In this context, it is instructive to take a closer look at what Mathew McCubbins and Thomas Schwartz (1984) have to say about techniques by which political authorities monitor subordinate agencies, narrow the information gap between policy-making principal and implementing agent, bring about adjustments in implementation, and secure a closer alignment between political intentions and bureaucratic implementation. McCubbins and Schwartz distinguish between two different strategies for the monitoring and political control of implementing agencies. First, "police-patrol oversight" is established at the initiative of the political principal. Such oversight is the pro-active, top-down, and direct approach to monitoring. The aim is to assess that implementation is congruent with political objectives. The second approach to the control of the bureaucratic implementation of policy-decisions is the monitoring technique of "fire-alarm oversight." This is the more passive, indirect, and decentralized approach to implementation monitoring. The political principal who conducts fire-alarm oversight establishes "a system of rules, procedures, and informal practices that enable [a political substitute] to examine administrative decisions [and] to seek remedies from agencies [...] itself" (McCubbins and Schwartz 1984: 166). Fire-alarm oversight is control of implementing agent using a political substitute or deputy.

Concerning caveats, police-patrol oversight implies that national political authorities intervene in the operational decision-making process to make sure that certain or all military operations are executed pending on the expressed approval from political authorities. In the language of the military, this implementation-controlling mechanism of "positive command" has the advantage of establishing tighter political control and reducing the risk of slippage between intentions and actions. This direct approach to political control of military implementation is attractive for political principals who fear that the use of force easily might escalate the conflict and change the nature of the mission itself—so-called mission creep. If political principals are preoccupied with avoiding mission creep and goal slippage for reasons of prospects-related risk reluctance, we thus expect the government to be inclined to apply restrictive caveats on the military contingent using the assertive mechanism of positive command.

Fire-alarm oversight is the preferred approach to national control over military implementation when tight and direct political control becomes impractical or too costly in time and effort for political principals. Drawn to tightly, restrictive caveats on the use of force and delegation of authority could also restrict action and limit operational initiatives that make military units vulnerable to attack, and might reduce their flexibility to handle sudden changes on the battlefield (Sagan 1991: 444).

In the context of caveats and NATO coalition operations, oversight by alarm will include the more extensive delegation of authority to a national military representative at different levels in the multinational chain of command. Recall from Chapters 3 and 4 that national officers assigned to coalition staff have been delegated the discretion to allow specified actions unless explicitly negated by countermanding orders from political authorities. In military terms, this is "command by negation." The national staff officers are placed to oversee military planning processes in the coalition and to assess what missions the national contingent be assigned. If a mission does not satisfy the criteria set by national political authorities, the military representatives have the delegated authority to refuse it against the orders of the Force Commander (Auerswald and Saideman 2014: 5–6).

If political decision-makers, for reasons of prospects-related risk willingness, are less concerned with the misplaced use of military force, and more concerned with providing military forces enough operational flexibility to ensure military efficiency, we expect the government to delegate authority to contingent commander to make decisions on the use of force, unless negated by countermanding orders from the government.

A main line of reasoning so far has been that the inclination of political principals to apply caveats on force contributions is related to prospects-conditioned risk assessment. Risk averse principals are likely to apply restrictive caveats in the form of positive command (a police-patrol oversight mechanism), and thus restrict delegation of authority to military to make own decisions on when to take on a mission. Risk-accepting principals, on the other hand, are more willing to leave decisions on the implementation of military operations to the military only limited by the boundaries of discretion and the countermanding orders (command by negation) from political authorities (a fire-alarm oversight mechanism). Hence, caveats based on the mechanism of positive command are likely to be more restrictive; caveats based on command by negation, considerably less so.

Note that that in the politics of implementation caveats are relevant mainly in terms of whether, how far, and on what terms the government is willing to delegate political authority to national contingent making decisions on the use of force. Less relevant is the regulatory dimension of caveats related to the use of force deviating from the common RoE. This begs the question as to how we can recognize a "no caveats applied" situation when we look at it? More precisely, what does a negative observation looks like on the operational dimension of caveats defined as the delegation of authority to make decisions on the use of force?

The answer is that, in the politics of implementation, a complete lack of caveats is observable in the utter delegation of military implementation to the military within the limits of the national RoE (restrictive or permissive, if any). Neither positive command nor command by negation applied to the contingent. There will be no pro-active monitoring mechanisms. The political principal relies entirely on the daily and weekly reports from the military agent without much questioning.

Why then would a contributing government decide to apply no caveats (full delegation of authority), or very limited restrictive caveats in the bureaucratic politics of implementation? First, the principal's risk acceptance might be extremely high because of prospects for huge losses (Levy 1996, 1997). When in dire straits, we may expect the principal to refrain from any political intervention which might hurt military efficiency. On the regulatory dimension of caveats, the principal would likely favor permissive caveats that are even more robust as to use of force than the coalition RoE allows for. The political principal may even want to add more troops to its military contribution.

Second, the full delegation of authority to the implementing agent may be due to the principal's need to establish plausible denial in case the mission backfires, politically or operationally. When mistakes are made and questions of accountability asked, political decision-makers who have refrained from micro-management in areas outside their expertise may avoid political costs by blaming bad judgment on military commanders.

Finally, McCubbins and Schwartz (1984) emphasize that monitoring of military implementation is costly in time, logistics, and human resources. Resource limitations are likely to influence the balance all coalition members must strike between implementation effectiveness (the extent to which military implementation fulfills political intentions) and military efficiency (the extent to which military force is utilized to the full). The first favors political control measures—including caveats, the latter favors a free hand to fight the war on military premises. Monitoring requires time and professional skills. However, key policy-makers have limited time. Moreover, without the necessary military skills, the intervention of political principals may negatively impact the efficiency of military decision-making processes (Feaver 2003: 100–101).

Frost-Nielsen argues that for the less resourceful members of the coalition, the failure to monitor and politically control the military implementation using positive command or command by negation may not be one of choice, but rather the lack of means. If so, what may look like a decision not to apply caveats is rather a lack of capacity to politically control own forces. However, the lack of capacity is a completely different phenomenon from caveats, and thus outside the boundaries of our current research interest.

POINTING THE ARGUMENTS: EMPIRICAL PROPOSITIONS

National reservations on the use of force may solve problems related to political principals' need to control that implementation adhere to the political intentions justifying coalition participation. Whether and to what extent the need for political control leads to the application of caveats, may depend on several factors, according to the literature reviewed and the deductive reasoning explained above. The main challenge facing the political principal is how to balance concerns for implementation effectiveness with implementation efficiency (Coletta and Feaver 2006; Ruffa et al. 2013: 325; Sagan 1991; Van Bezooijen and Kramer 2014).

The optimal trade-off in this dilemma depends on several factors and mechanisms discussed. The main take-home message is nevertheless that caveats is an instrument potentially capable of facilitating the fine-tuning of the trade-off between contradicting concerns also in the politics of implementation. A final observation is that in the realm of implementation, caveats are relevant mainly in terms of how far and on what terms the government is willing to delegate political authority to make decisions on the use of force, whether using "positive command" or "command by negation." Less relevant is the operationalization of national reservations on the use of force as deviations from the common RoE. We offer the following deductions as points of departure for empirical research:

"Politics of implementation" hypotheses (PI-H1–14): Everything else being equal, we may consider that;

PI-H1   *"Principal's dilemma" hypothesis:* Political efforts to maximize military effectiveness using restrictive caveats, is likely to reduce military efficiency.

PI-H2   *"No-arm-tied-on-the-back" hypothesis I:* To the extent the military agent act as veto-player in the planning and decision-making on the participation in the coalition, this decreases the political principal's inclination to apply restrictive caveats on the national contingent.

PI-H3   *"No-arm-tied-on-the-back" hypothesis II:* To the extent the political principal applies restrictive caveats to national contingent, the military agent is inclined to apply a flexible interpretation of the caveats imposed.

PI-H4 *"Too strong for comfort" hypothesis:* The more robust the coalition force, the more inclined are political principals to apply restrictive caveats on the national contingent.

PI-H5 *"Too close for comfort" hypothesis:* The more integrated into other nations' forces the national contingent becomes, the more inclined are political principals to apply restrictive caveats.

PI-H6 *"Tolerance for political risk" hypothesis I:* Political principals with a risk averse mindset have a strong incentive to monitor and apply national reservations on the use of force.

PI-H7 *"Tolerance for political risk" hypothesis II:* The more willing political principals are to accept the potential repercussions from domestic politics, the less inclined they are to apply restrictive caveats.

PI-H8 *"Tolerance for political risk" hypothesis III:* The more willing political principals are to accept the risk of being singled out for criticism by allies, the more inclined they are to offer a limited military contribution or apply restrictive caveats.

PI-H9 *"Prospects for gain and risk acceptance" hypothesis:* The more the political principal frames the participation in the coalition in terms of prospect for gains, the less risk the principal is willing to accept, and thus more inclined to apply restrictive caveats.

PI-H10 *"Prospects for loss and risk acceptance" hypothesis:* The more the political principal frames the participation in the coalition force in terms of prospect for loss, the more risk the principal is willing to accept, and thus less inclined to apply restrictive caveats.

PI-H11 *"Police-patrol oversight" hypothesis:* Political principals preoccupied with avoiding mission creep and goal-slippage are inclined to apply restrictive caveats on the military contingent using the assertive mechanism of "positive command."

PI-H12 *"Fire-alarm oversight" hypothesis:* Political principals preoccupied with contingent's military flexibility and efficiency are inclined to delegate authority to the military to make decisions on the use of force through the mechanism of "command by negation," to the extent applying restrictive caveats at all.

PI-H13   *"Crisis-management"* *hypothesis:* Political principals accepting
          high levels of risk due to prospects for huge losses are inclined
          to remove restrictive caveats in the chain of command, apply
          permissive caveats on the use of force, and/or strengthen the
          military contribution with more troops.

PI-H14   *"Political insurance"* *hypothesis:* Political principals in need of
          establishing "plausible deniability" in case the mission back-
          fires are inclined to delegate full authority to military agents
          to decide how to execute the military operation.

Having theorized how bureaucratic politics of implementation in the
realm of civil–military relations may influence political decision-makers'
inclination to delegate authority to the military to make decisions on
the use of force, it is crucial to emphasize in the closing section that the
multi-level approach of FPA is not primarily about studying each level of
analysis and phase of decision-making in separation. The separate analyses of alliance politics, domestic politics, and the politics of implementation are merely a theoretical and empirical preparation for the final study
of how factors at different levels of analyses *interact* in producing variation in the use of caveats in foreign policy and between contributing
states in coalition operations. Whether this interaction takes the shape of
multiple paths of causation is, of course, an empirical question.

In such a multi-level *synthesis* of partial analyses, the emphasis is on
pulling together how enabling and limiting structural factors, as well
as agency-related intentions and narratives theorized above, interact in
"pushing the envelope" on the principal and practical questions of war
participation through different phases of the decision-making and implementing processes. As mentioned in previous chapters, FPA is not the
only approach in Political Science making use of several levels of analyses as a key defining epistemological attribute. Robert Putnam developed
the "Two-level game" approach (1988) to make International Relations
(IR) scholars pay more attention to the interaction between domestic
and international politics in explaining outcomes of international negotiations. The article has since become a standard text in IR.

Closer to home, Joseph Lepgold and Brent Sterling (2000) draw
upon the dual political environments of foreign policy-making by explicitly distinguishing between two separate decision-making risks as related
to the use of armed force: that related to the risk of repercussions from

international politics, and that related to the risk of losing domestic support for coalition participation.

Finally, also Samuel Huntington addresses several levels of analyses. In his seminal study, The Soldier and the State (1957), Huntington argues that civil–military relations very much are shaped by the political decision-makers perception of external threats, and domestic factors related to the structural and ideational attributes of state and society. Such framed, Huntington's approach is FPA in much, except the name.

REFERENCES

Auerswald, D. P., & Saideman, S. M. (2014). *NATO in Afghanistan: Fighting Together, Fighting Alone*. Princeton, NJ: Princeton University Press.

Boettcher, I. I. I., & William, A. (2004). Military Intervention Decision Regarding Humanitarian Crises: Framing Induced Risk Behavior. *Journal of Conflict Resolution, 48*(3), 331–355.

Chopra, J., Eknes, Å., & Nordbø, T. (1995). *Peacekeeping and Multinational Operations*. Oslo: Norwegian Institute of International Affairs.

Clarke, M. (1996). Foreign Policy Analysis: A Theoretical Guide. In S. Stavridis & C. Hill (Eds.), *Domestic Sources of Foreign Policies: Western European Reactions to the Falkland Conflict* (pp. 19–39). Oxford: Berg.

Cochran, S. T. (2014). The Civil–Military Divide in Protracted Small War: An Alternative View of Military Leadership Preferences and War Termination. *Armed Forces and Society, 40*(1), 71–95.

Cohen, E. A., & Gooch, J. (2006). *Military Misfortunes—The Anatomy of Failure in War*. New York, NY: Free Press.

Coletta, D. (2013). Principal-Agent Theory in Complex Operations. *Small Wars & Insurgencies, 24*(2), 306–321.

Coletta, D., & Feaver, P. D. (2006). Civilian Monitoring of U.S. Military Operations in the Information Age. *Armed Forces and Society, 33*(1), 106–126.

Cronin, B. (2013). Reckless Endangerment Warfare: Civilian Casualties and the Collateral Damage Expectation in International Humanitarian Law. *Journal of Peace Research, 50*(2), 175–187.

Feaver, P. D. (1998). Crisis as Shirking: An Agency Theory Explanation of the Souring of American Civil–Military Relation. *Armed Forces and Society, 24*(3), 407–434.

Feaver, P. D. (1999). Civil–Military Relations. *Annual Review of Political Science, 2*(June), 211–241.

Feaver, P. D. (2003). *Armed Servants. Agency, Oversight, and Civil–Military Relations*. Cambridge, MA: Harvard University Press.

Frost-Nielsen, P. M. (2016). *Betingede forpliktelser. Nasjonale reservasjoner i militære koalisjonsoperasjoner*. Ph.D. Dissertation in Political Science. Trondheim: Department of Sociology and Political Science, Norwegian University of Science and Technology (NTNU).

Frost-Nielsen, P. M. (2017). Conditional Commitments: Why States Use Caveats to Reserve Their Efforts in Military Coalition Operations. *Contemporary Security Policy, 38*(3), 371–397.

George, A. L. (1991). The Tension Between "Military Logic" and Requirements of Diplomacy in Crisis Management. In A. L. George (Ed.), *Avoiding War—Problems of Crisis Management* (pp. 124–143). Boulder, CO: Westview Press.

Goodpaster, A. J., & Huntington, S. P. (1977). *Civil-Military Relations*. Omaha, NE: American Enterprise Institute.

Greentree, T. R. (2013). Bureaucracy Does Its Thing: US Performance and the Institutional Dimension of Strategy in Afghanistan. *Journal of Strategic Studies, 36*(3), 325–356.

Hasik, J. (2014). 'Outside Their Expertise': The Implications of Field Manual 3–24 for the Professional Military Education of Non-commissioned Officers. *Small Wars & Insurgencies, 25*(5–6), 1055–1062.

Huntington, S. P. (1957). *The Soldier and the State: The Theory and Politics of Civil–Military Relations*. Cambridge, MA: Belknap Press.

Ingesson, T. (2016). *The Politics of Combat: The Political and Strategic Impact of Tactical-Level Subcultures, 1939–1995*. Lund: Ph.D. Dissertation, Faculty of Social Sciences and Department of Political Science, Lund University. http://portal.research.lu.se/ws/files/7253766/Tony_Ingesson_Politics_of_Combat.pdf.

Ingesson, T. (2017, September 20). *Trigger-Happy, Autonomous, and Disobedient: Nordbat 2 and Mission Command in Bosnia*. The Strategy Bridge. https://thestrategybridge.org/the-bridge/2017/9/20/trigger-happy-autonomous-and-disobedient-nordbat-2-and-mission-command-in-bosnia.

Kahl, C. H. (2007). In the Crossfire or the Crosshairs? Norms, Civilian Casualties, and U.S. Conduct in Iraq. *International Security, 32*(1): 7–46.

Kier, E. (1997). *Imagining War: French and British Military Doctrine Between the Wars*. Princeton, NJ: Princeton University Press.

King, A. (2006). The Word of Command—Communication and Cohesion in the Military. *Armed Forces and Society, 32*(4), 493–512.

Knudsen, E., & Klingenberg, S. (2013). *Cooperating in War—Coalition Warfare in Afghanistan*. Copenhagen: Forsvarsakademiet.

Kowert, P. A., & Hermann, M. G. (1997). Who Takes Risks? Daring Caution in Foreign Policy Making. *Journal of Conflict Resolution, 41*(5), 611–637.

Lepgold, J., & Sterling, B. L. (2000). When Do States Fight Limited Wars? Political Risk, Policy Risk, and Policy Choice. *Security Studies, 9*(4), 127–166.

Levy, J. S. (1996). Loss Aversion, Framing, and Bargaining: The Implications of Prospects Theory for International Conflict. *International Political Science Review, 17*(2), 179–195.

Levy, J. S. (1997). Prospect Theory, Rational Choice, and International Relations. *International Studies Quarterly, 41*(1), 87–112.

McCubbins, M. D., & Schwartz, T. (1984). Congressional Oversight Overlooked: Police Patrols Versus Fire Alarms. *American Journal of Political Science, 28*(1), 165–179.

Meyer, T. (2013). Flipping the Switch: Combat, State Building, and Junior Officers in Iraq and Afghanistan. *Security Studies, 22*(2), 222–258.

Miller, G. J. (2005). The Political Evolution of Principal-Agent Models. *Annual Review of Political Science, 8*(2), 203–225.

Noetzel, T., & Rid, T. (2009). Germany's Options in Afghanistan. *Survival, 51*(5), 71–90.

Osiel, M. J. (1999). *Obeying Orders: Atrocities, Military Discipline, and the Law of War.* New Brunswick, NJ: Transaction Publishers.

Posen, B. R. (1984). *The Source of Military Doctrine—France, Britain, and Germany Between the World Wars.* Ithaca, NY: Cornell University Press.

Rauchhaus, R. W. (2009). Principal-Agent Problems in Humanitarian Intervention: Moral Hazard, Adverse Selection, and the Commitment Dilemma. *International Studies Quarterly, 53*(4), 871–884.

Rosen, S. P. (1995). Military Effectiveness: Why Society Matters. *International Security, 19*(4), 5–31.

Ruffa, C. (2013). The Long and Winding Road to Success? Unit Peace Operation Effectiveness and Its Effect on Mission Success. *Defense and Security Analysis, 29*(2), 128–140.

Ruffa, C. (2014). What Peacekeepers Think and Do: An Exploratory Study of French, Ghanian, Italian, and South Korean Armies in the United Nations Interim Force in Lebanon. *Armed Forces and Society, 40*(2), 199–225.

Ruffa, C., Dandeker, C., & Vennesson, P. (2013). Soldiers Drawn Into Politics? The Influence of Tactics in Civil–Military Relations. *Small Wars & Insurgencies, 24*(2), 322–334.

Sagan, S. D. (1991). Rules of Engagement. In A. L. George (Ed.), *Avoiding War—Problems of Crisis Management* (pp. 443–470). Boulder: Westview Press.

Scheipers, S. (2014). Counterinsurgency or Irregular Warfare? Historiography and the Study of 'Small Wars'. *Small Wars & Insurgencies, 25*(5–6), 879–899.

Snyder, J. (1984). Civil–Military Relations and the Cult of the Offensive 1914 and 1984. *International Security, 9*(1), 108–146.

Soeters, J., & Manigart, P. (2008). Introduction. In J. Soeters & P. Manigart (Eds.), *Military Cooperation in Multinational Peace Operations—Managing Cultural Diversity and Crisis Response.* Oxon: Routledge.

Soeters, J., van Fenema, P. C., & Beeres, R. (2010). Introducing Military Organizations. In J. Soeters, P. C. van Fenema, & R. Beers (Eds.), *Managing Military Organizations—Theory and Practice* (pp. 1–14). London: Routledge.

Sookermany, A. M. (2012). What Is a Skillful Soldier? An Epistemological Foundation for Understanding Military Skill Acquisition in (Post) Modernized Armed Forces. *Armed Forces and Society, 38*(4), 582–603.

Van Bezooijen, B., & Kramer, E.-H. (2014). Mission Command in the Information Age: A Normal Accidents Perspective on Networked Military Operations. *Journal of Strategic Studies, 38*(4), 445–466.

Von Hippel, K. (2000). *Democracy by Force. US Military Intervention in the Post-Cold War World.* Cambridge: Cambridge University Press.

Weber, M. (2010). *Makt og byråkrati.* Oslo: Gyldendal Norsk Forlag.

PART V

Researching Caveats

CHAPTER 11

Considerations and Recommendations for the Gathering and Analyses of Data

What advice can be offered to scholars adopting the multi-level and decision-making FPA approach for the empirical research of coalition participation in general and the reservations on the use of force in particular? Methodological implications of the research program have been alluded to in several chapters, but not been dealt with systematically. In the present chapter, we first discuss issues related to the collection of data, and second the question of what methodological approaches to empirical analysis may give more epistemological traction in the study of the foreign politics of national reservations on the use of force. As to the first task of gathering data, we need to address not only the empirical implications of the analytical framework offered. As crucial, we need to point out steps to rectify some existing limitations in the caveats-literature and find ways of dealing with the fact that reliable data on national reservations on the use of force can be hard to get at because access to caveats-policies may be restricted for concerns of national security. Concerning the second task of analyzing data, we discuss how single and comparable-cases designs be applied for both theory-developing and theory-testing purposes in our particular field of research.

© The Author(s) 2019
G. Fermann, *Coping with Caveats in Coalition Warfare,*
https://doi.org/10.1007/978-3-319-92519-6_11

SYSTEMATIC DATA A PRECONDITION FOR THE CREATION OF GENERIC KNOWLEDGE

The scholarly literature on the politics of caveats suffers from two limitations related to data. One is the lack of anything close to a comprehensive database on the use of caveats in coalition operations in, say, the post-Cold War era. Expanding the scope to include even previous coalition forces, and to cover variables potentially explaining patterns of caveats, the scarcity of data becomes even more apparent.

Second, as to the data sets already compiled, the main problem is that data originating from different studies are unlikely to be comparable in any strict sense of the term. This is, as indicated, because different scholars define the concept of caveats differently, or operationalize the variable in only partly overlapping ways. A common conceptualization of a phenomenon is a necessary condition for the creation of systematic data, which, in turn, is a precondition also for the construction of general knowledge on caveats regarding distribution and explanation. Thus without scholars relating to a common-core conception of caveats, we will not be able to build a systematic database on the phenomenon of caveats that can feed comparative and large-N research, and thus produce generic knowledge in a systematic sense.

We have defined caveats as national reservations on the use of force in coalition-contexts and suggested a three-dimensional operationalization of the caveats-variable as a basis on which to compile data systematically. Indeed, in a string of case studies, Per Marius Frost-Nielsen (2016, 2017) has shown how this compound conceptualization of caveats is capable of capturing and distinguishing between different kinds of reservations on the use of force. The suggested conceptualization of caveats is also capable of detecting and excluding patterns of operational behavior, which is due to imperfect coordination or lack of technical capacity that could otherwise be mistaken for politically motivated caveats.

As to the host of independent variables that may potentially explain patterns of caveats across nations and coalitions, we may avoid unproductive work by relating closely to well-established definitions where they exist in the relevant literature. Scholars on coalition dynamics should, for instance, avoid developing their particular conception of some independent variable for caveats-research. Instead, we need to go to the alliance politics literature to see how compound concepts such as fear of entrapment and fear of abandonment are operationalized (see empirical propositions AP-H1 through AP-H3), and to the huge literature on Comparative Politics to be meticulous about how the concept of, say, "cosmopolitan outlook" (e.g., empirical proposition DGP-H12) be measured.

When using established operationalization of independent variables, we should be very explicit about it. Not only will this facilitate comparability within our particular field of research. Scholars in other areas of political research may readily use our research findings for their purposes, thus contributing to cumulative research in a much broader sense. We may fuse smaller modules of knowledge into larger bodies of knowledge. That is, to the extent the components connecting them—validly operationalized concepts, in particular—actually fit together ontologically. Again, common conceptions and well-argued operationalization of relevant phenomena are preconditions for the harmonization and fusion of knowledge across data sets.

GATHERING DATA IN A POLICY DOMAIN PERVADED BY NATIONAL SECURITY CONCERNS

While the conceptual and analytical framework argued in the present study provides clues to what data to gather and analyze, we still need to find ways to deal with the fact that not all relevant information on caveats are readily available. In particular, this is likely to include nation- and coalition-specific information on RoE, command-structures, bilateral force agreements, and burden-sharing settlements in NATO. It may also include information on some of the independent variables suggested for further scrutiny in Chapters 8–10.

In a policy-area pervaded by concerns about national security and political accountability, we should expect some information to be classified or hard to find. We should also expect to see variations in how willing states are to offer information on coalition participation and in particular on national RoE. For instance, the United States tends to be more willing to make available information on RoE than the United Kingdom. We have also experienced that information which is hard to retrieve from national authorities, may be available from coalition sources.

Nevertheless, political and military sensitivity provides an incentive to control information flows and restrict access to caveats-relevant information. For instance, Force Commanders most often do not want complete information on national reservations on the use of force, and coalition RoE broadcasted to the enemy. Governments applying restrictive caveats do not want to see their contribution belittled in the public eye, even if restrictive caveats may be the reason why a coalition government can agree upon coalition participation in the first place. In case of collateral damage or the loss of own troops, governments applying permissive caveats would prefer to avoid the impression that dire consequences are the result of lack of military restraint, or due to reckless operational behavior in the field. A much more trivial barrier to caveats fact-finding is that not all relevant information is available in the English language, the contemporary *lingua franca* of the world.

Also note that information gathering within the approach of Foreign Policy Analysis (FPA) is considerably more demanding than just collecting data on caveats and some framework condition, say, whether the coalition government in question is Center-Left or Center-Right. FPA also implies an empirical ambition to trace precisely how global and domestic framework conditions are interpreted, and acted upon in foreign policy-making and implementing processes. This may include the attempt to map what considerations were made in the assessment of scope for political maneuvering (SPM), in the deciding on goals and preferences, in the calibration of policy instruments, and in the structuring of the interface between political decision-making and military implementation (Fermann 2013: 89–139). This kind of information is rarely available in some pre-made generic database, but in high demand among FPA researchers conducting a process-tracing analysis of policy decisions and implementation behavior (Bennett and Checkel 2015).

Unlocking the proverbial "black box" of foreign policy-making by applying approaches such as governmental/bureaucratic politics, organizational study, political psychology, and the study of public policy implementation is quintessential FPA (Hudson 2007; Allison and Zelikow 1999). Decision-making study allows us to understand better the *causal chain*—the string of mechanisms—mediating the relationship between some external framework factors (international or domestic) on the one hand and national coalition-behavior on the other. IR and CP inspired studies preoccupied with systemic and domestic variables for the benefit of high-N generalizing tend to ignore, assume, or only theoretically argue such process mechanisms.

However, the cost of filling this intermediate epistemological gap with empirical research is, as indicated, a reduction in parsimony. We pay this considerable research-bill in part at the stage of analyzing data, but even more in the initial stage of gathering relevant decision-making process information. The more crucial it is to look at ways to come around obstacles to data gathering in the particular field of coalition participation. What can scholars do to circumvent political and military incentives to protect caveats-relevant information against exposure?

Aside from being equipped with a precise conception of caveats that render possible the recognizing of the phenomenon when observed, we have found it particularly useful to include political scientists with a military background in the research team, and to establish research cooperation with military schools. The military academic and the political scientist working from or in close cooperation with a military school or agency are likely to have access to caveats-relevant information that is not nearly as accessible for the less connected civilian researcher. In particular, there may be a promising sweet spot for information-gathering open for the militarily connected researcher in-between what can be found in open sources and outlets for research (white papers, news reporting, academic journals, and books), and what is restricted/classified information or information regulated by some nondisclosure agreement. This spot is nonclassified and caveats-relevant information embedded in networks and hierarchies of policy-makers, military planners and operational command (Stake 1995: 49–68).

Well-placed researchers and military academics can get access to data in the shape of unclassified documents, through interviews and even participatory observation. Participatory observation may imply the researcher be deployed in the chain of military command (Kawulich 2005). As participatory observation in the field also counts the researcher's caveats-relevant experience in military service prior to engaging in research. The officer-turned-scholar is in a position to utilize military and institutional local knowledge and professional skills to find the most abundant sources of information, get access to key military personnel, ask the door-opening questions, and to better interpret, contextualize and follow-up on the feedback they receive from their interviewees.

By getting access to the higher echelons of military command, the military academic may also retrieve crucial information on what is going on at the interface between political decision-makers and top military brass (Auerswald and Saideman 2014; Henriksen 2007). Such information gathering may reveal the existence of informal caveats of the restrictive or permissive kind. We have also found that information gathered from mid- and lower-level officers are valuable in revealing the existence of national RoE deviating from coalition RoE, the actual behavior of national gate-keepers in the chain of military command, and particular conditions on the use of force codified in national force agreements and burden-sharing settlements in NATO (Frost-Nielsen 2016). Finally, the military scholar is likely to have a keener sense of where to draw the line to the use of sensitive information in publicly available research than their civilian counterpart. The optimal balancing of access to sensitive information and the liberal use of the same information in research is likely to benefit from previous experience in researching in a security-sensitive environment.

No matter how well equipped the research team is for the task of gathering information, caveats-relevant data on decision-making and implementing processes is unlikely to all come in the form of hard evidence, the proverbial "smoking gun." Data often surfaces as circumstantial evidence, indications that a state's decision on caveats was such-and-such, for this-or-that reason. Circumstantial evidence provided through unconfirmed statements, rumors, and perspective-dependent sources is a legitimate empirical part of the picture we form about hard-to-establish framework factors; this-or-that decision-maker; certain kind of priorities; and particular assessments of means–end relationships made behind closed doors. While having a sharp eye on the reliability of the sources and the validity of the information gathered, scholars make use of all kinds of retrievable data to the extent it can be justified by the research questions posed, the approach applied and the operational definitions decided upon. Along with theoretical insights, hard and circumstantial evidence are systematically triangulated to establish the relevant facts, and subsequently, to argue causal relationships. The latter requires the mobilization of proper methods for empirical analysis.

Empirical Analyses and the Foreign Politics of Caveats

The final contribution of the study is to indicate some promising methodological designs for the analyses of data while taking into consideration both the present state of political research on caveats and the methodological implications following from the choice of making the FPA approach a crucial element of the empirical research program. Generally, the aim of explanatory empirical analysis is to unravel the conditions explaining some phenomenon, by inferring credible causal relationships based on the best available data and a convincing set of theoretical arguments. In particular, we have offered an analytical framework as a point of departure to identify enabling and limiting structural conditions as well as agency-related factors to explain coalition participation patterns and caveats behavior in particular. The question remains as to what empirical research methods are likely to give more epistemological traction in the empirical study of the foreign politics of caveats at the present stage of research.

Ways of Knowing

In Social Science, causal explanations need to fulfill four fundamental criteria to make credible claims to explanatory knowledge. To assert that some phenomenon is causally conditioned (explained) by some other phenomenon, the relationship between cause (X) and effect (Y) need to be both (i) theoretically argued in terms of what mechanism is bringing about the causal relationship, and (ii) empirically supported in terms of reliable data. Furthermore, (iii) the cause must appear before the effect in time, thus establishing the direction of the empirical relationship. Finally, (iv) to safeguard against spuriousness, the observed relationship between cause (X) and effect (Y) must survive the scientific control for other well-argued conditions (say, Z1, Z2, and Z3) by means of some methodology capable of keeping control variables constant (Møller 2015: 89–90). To this, we may add a fifth requirement: The causal path of mechanisms mediating and contributing to the causal relationship between an X and a Y should be empirically demonstrated and not merely theoretically argued or assumed.

Depending on the research question posed, the particular unit of analysis, the number of observations available, and the measurement level of data (nominal, ordinal, interval, ratio), a wide range of methodological techniques for the scientific control for third variables are available for Social Science empirical research. On a scale ranging from "hard" to "soft" scientific control, the repertoire of methods for the construction of causal inference includes (i) the laboratory-controlled experiment; (ii) a host of statistical techniques relying on a large number of observations; (iii) comparative methods; and (iv) single case-study designs. Within each one of these methodological approaches, multiple techniques and research practices have been developed (e.g., Bartlett and Vavrus 2017; Byrne and Ragin 2013; Moses and Knutsen 2012; Outhwaite and Turner 2011; Brady and Collier 2010; Rihoux and Ragin 2009; Johnson et al. 2007; Bryman 2001; King et al. 1994; Ragin and Becker 1992). Add to this the hermeneutic interpretive approaches which have been explored to make sense of the social world (e.g., Moses and Knutsen 2012: 169–298; Kögler 2011: 363–383; Bauman 1978), and the arsenal of epistemological and methodological approaches available to shed light on social and political phenomena is both impressing and somewhat intimidating.

Fortunately, for our particular purposes, there is no need to apply the comprehensive textbook approach to the review of research methods. A way of narrowing down our current recommendation of research designs on the politics of caveats is to ask (i) what gaps in our knowledge need filling in the present less-than-saturated literature on the politics of caveats, and (ii) what kinds of empirical analyses are called for given the framework for analysis elaborated on in the previous? From the outset, we rule out two major methodological approaches.

The laboratory-controlled experimental method is less relevant for our purposes because the phenomenon of coalition behavior does not satisfy the strict requirements of this methodology (Schneider 2007: 173; Bryman 2001: 32–40). It is unlikely that we can isolate the conditions for the application of caveats with any glass walls of the experimental laboratory.

Furthermore, applying statistical techniques for the control of third variables is premature since no systematic and comprehensive database on cross-coalition and cross-country caveats exists that yet allows for large-N studies of the politics of caveats. Until scholarly consensus on how to measure caveats is established, the fulfillment of the generic ambition driving statistical large-N studies belongs further down the road of empirical research.

The question remains then as to what empirical research methods are likely to give more epistemological traction in the empirical study of the politics of caveats at the present stage of research. *Both the exploratory phase of political caveats-research and the foreign policy-making framework for analysis argued indicate that single and comparable-cases designs are the most promising methodological approaches to empirical research on caveats at this time.* Indeed, the data-intensive design of the case study is very much the methodological mirror image of the epistemological ambition of the FPA approach to connecting causally the empirical and epistemological dots between structural framework factors and caveats behavior. Empirically, we recognize the dots as foreign policy decision-making and implementing processes. Epistemologically, we conceptualize the dots as causal paths and mechanisms. However, the "case-study design" is not a singular method, but an epistemological approach covering a wide range of techniques, which may be scrutinized to serve different, but complementary epistemological purposes in the research process.

A Case of What? The Generic Research Ambition

What then is the essence of the case study, and what distinguishes the historian's idiographic approach to the case study from the broader, nomothetic research ambition of the Social Scientist? There is no unified understanding of a "case," a "case study," and the "case-study methodology" across disciplines and subfields except a commitment to make sense of rich and deep data (Platt 2007: 115–116). This data-intensive and context-rich approach to scientific inquiry do restrict the number of cases the researcher is capable of analyzing. According to Charles C. Ragin (2000: 25), case-study methodologies are applicable in the single case and small and intermediate-size multiple case studies, up to some 50 cases.

Jack S. Levy (2008: 2) sees the case study as "an attempt to understand and interpret a spatially and temporally bounded set of events," or phenomena. Thus, a case is not merely a unique observation, but considered "an instance of [a theoretically defined] class of events" (George and Bennett 2005: 17). Similarly, John Gerring identifies the case study as "the intensive study of a single case where the purpose of that study is to [...] shed light on a larger class of cases" (2007: 13). This epistemological ambition of generalization exceeds but does not contradict or make obsolescent the study of a case as a singular historical subject.

Such understood, we consider the present research program a case study in its own right. However, our research program on the foreign politics of caveats is not a case study limited to the mainly descriptive sense of the "a-theoretical" approach of Arendt Lijphart (1971: 689) or the "configurative-ideographic" approach of Harry Eckstein (1975: 96). Neither is it a particular case study in the theory-guiding sense of the "interpretive," "disciplined-configurative," and the "case-explaining" approach as elaborated by Lijphart (1971: 691), Eckstein (1975: 99) and Stephen Van Evera (1997: 89–91) respectively. This is because these particular case-study designs are mainly preoccupied with shedding light on the particular and possibly unique circumstances relating to a discrete case/event (ideographic). By definition, the epistemological ambition of a Social Science empirical research program such as ours is generic (nomothetic) (Hall 2007: 83–85). Indeed, in using the historical material, Social Scientists are primarily engaged in the explanation of generic phenomena, whether for theory-generating or hypotheses-testing purposes.

The question remains, however, how do we *transcend* from the study of the singular event to the study of a generic phenomenon in case-study research? In other words, how do we go about "transforming descriptive explanations into analytical explanations" (George and Bennett 2005: 92)? The generic transition starts by asking "what broader phenomenon is this instance a case of" (Platt 2007: 104), thus shifting the focus from the historian's idiographic task of understanding a singular case, to the nomothetic and theory-informed project of shedding light on a class of cases (Levy 2008: 4–5; Gerring 2007: 13, 19–20; Ragin and Becker 1992).[1] As discussed in Chapter 2, the answer to the question of what broader class of phenomenon a case belongs to, not only determines the empirical boundaries (the population) for case selection in single and multiple case studies.[2] Equally crucial, the answer also provides critical clues to what analytical framework will inspire the formulation of research questions and from what bodies of theory explanatory hypotheses be deduced (Thomas 2011: 512).

Indeed, Michel Wieviorka argues that what identifies a case as a generic phenomenon, is not the sharing of certain characteristics as such. What makes cases belong to the same class of phenomenon is rather the analytical framework chosen by the researcher to make sense of the phenomenon (1992: 159–160). To ask what a case, is "a case of" is thus not only a declaration of a generic research ambition but also an invitation to define the scope and identify the analytical content of the generic phenomenon under study. In the words of Jennifer Platt:

---

[1] Robert E. Stake makes a similar, if not identical distinction in assigning the term "intrinsic" for case study undertaken "because that case, in particular, is of interest" (idiographic), and "instrumental" case study to indicate that the case/s "throw light on matters beyond the case/s studied" (nomothetic) (1995: xi).

[2] We will subsequently discuss case selection as a means of making causal inferences in single and multiple case studies.

It follows that the issue of whether one can generalize from case study is an issue not of the number of cases studied but the adequacy of the theory in relation to which it is interpreted, and the cogency of its theoretical interpretation against a background of knowledge of other cases. (Platt 2007: 113)

In our context, the *subject* of the case study is the particular caveats applied (or not) by a specific state in a particular coalition context with this-or-that impact on the mission. Here, the epistemological challenge is descriptive in the strictly idiographic sense. As to the *object* of the case study, this follows from the analytical framework chosen to make sense of caveats: Applying the FPA approach and middle-range theories at different levels of analyses, students of the foreign politics of caveats take on the epistemological challenge of explaining caveats in the nomothetic sense, within the ontological boundaries of the analytical framework. For generic purposes, we will engage in the *description* of the particular subject of national reservations on the use of force, but predominantly as part of the process of gathering data with the aim of *explaining* caveats as a generic object.

A final point pertaining to the essence of the case study is Helen Simon's observation that the case study is not a method in and on itself. Rather, the case study is a *design frame* that incorporates a number of methods and approaches applied for different but complementary epistemological purposes, including but not limited to the empirical testing of hypotheses (2009: 443). While single and multiple case-study methodologies serve the researcher's more ultimate aim of assessing empirical propositions, the case-study design also involves approaches and practices contributing to conceptual clarification, theory building (induction), and hypotheses generation (deduction) (George and Bennett 2005: 74–83). Hence, case-study design addresses all the basic building blocks of an empirical research program as defined in the present study. Framed as a case study in its own right, what are the main contributions of the research program on the politics of caveats so far?

As to the elaboration of useful analytical constructs and the task of theory building, the initial step of the research program was to argue that caveats be treated as a generic phenomenon rather than merely a particular observation. To do this required extending the notion of caveats as an event-specific proper noun to caveats as a concept. In support of the generic ambition, we reviewed in Chapter 3 the historical record, operational practices, and the scholarly literature to develop a firmer ground to assess the conceptual delimitation of caveats as national reservations on the use of force in coalition warfare. Without recapitulating in detail, we operationalized caveats along three dimensions, capturing different facets of the phenomenon. This made it possible to both recognize the generic phenomenon of caveats when observing it and to distinguish between different kinds and blends of caveats.

The second step was to assess what caveats might precisely be a case of. We discussed several alternatives. Eventually, we decided to proceed with the notion that national reservations on the use of force is a foreign policy instrument the sovereign state has at its disposal when assessing how to respond to alliance partners' call for contributions to coalition operations. Thus evolved caveats from a particular subject frozen in time and space (idiographic) to an object of broader and more generic significance (nomothetic). That is, a policy instrument in the multi-level politics of foreign policy-making, and thus a negotiating card in burden-sharing debates between allies.

The third step was to ask what this initial wave of inductive inferences implied for the choice of analytical approach and subsequently for the choice of substantial theories (George and Bennett 2005: 109–124). Recall that that approaches and theories are interpretive frameworks used to inspire research questions and to argue causal relationships based on the particular direction of the approach, and the ontological content of theories on the nature of political systems and their structure, the identity and motivation of actors, and the relationship between actors. In Chapter 5, this line of reasoning led us to the doorstep of FPA, a multi-level approach and decision-making approach directing us toward several bodies of middle-range theory. To summarize, in FPA national reservations on the use of force in coalition warfare is framed as a foreign policy instrument, researched as actual decision-making and implementing processes, and theorized as to how these processes, are influenced by forces on the global, domestic, institutional, and individual decision-making levels of analyses. From the chosen bodies of theory, hypotheses are deduced and subsequently tested.

Note that the travel toward generic knowledge may start from the ground-up observation of some barely recognized empirical puzzle in need of conceptualization and theorizing to make sense prior to any further testing (induction), or from the imagining of analytical constructs used to discover and making sense of previously unrecognized phenomena and relationships (deduction). We are not limited to see this as a choice between two, disconnected one-way streets. Quite the opposite, the case-study design is particularly well equipped to combine inductive and deductive reasoning for the ultimate purpose of theory testing, thus connecting methodological approaches into an epistemological two-way street (Moses and Knutsen 2012: 143).

In the final two main sections of the Chapter, we show how the case-study designs of the *deviant case*, the *plausibility probe*, and *process tracing* be applied for both theory-developing and theory-testing purposes (Eckstein 1975: 104–108; Lijphart 1971: 691). The case-study design of process tracing is in part developed to test hypotheses relating to the intermediary causal chain connecting input factors and a policy outcome (Brady et al. 2010: 15–32; George and Bennett 2005: 214–215). However, we may also use process tracing for theory-building purposes (Pouliot 2015: 237–259). Of course, any hypotheses generated from an exploratory and theorizing case study will require testing against other cases not included in a preceding theory-building case study. Yet other case-study designs, such as John Stuart Mill's four comparable-cases methods, and Qualitative Comparative Analysis (QCA) are specialized to serve mainly hypotheses-testing purposes (Eckstein 1975: 108; Mill 2002 [1891]; Frendreis 1983: 255–272; Przeworsky and Teune 1970; Ragin 2009; Rihoux and Ragin 2009; Rihoux 2013). Of course, findings resulting from comparative hypotheses-testing procedures should feed into subsequent theory-adjusting efforts.

Case Study for Theory Development

Even though single case studies are applied for hypotheses-testing purposes, case studies are often cast in the less heroic role of a preliminary to theory-testing research. However, as noted by Jennifer Platt, hypotheses testing "cannot start without some sense of what the realities are that need to be accounted for" (Platt 2007: 106, 120). Indeed, Alexander L. George and Andrew Bennett argue that "improved historical explanations of individual cases are the foundation for drawing wider implications from case studies, as they are a necessary condition for any generalization beyond the case" (2005: 110). Kathleen E. Eisenhardt reasons that in the early stages of research, theory building from case-study research is most appropriate because "theory building from case studies does not require previous literature or prior empirical evidence" (1989: 548). Jack S. Levy further recommends the researcher to explore pilot case study to assess if more costly and time-consuming research in multiple case studies and statistical analysis is worthwhile (2008: 6).

Levy's recommendation speaks caution to the practice of executing large-N research without prior in-depth study at the single case level of promising independent variables and possible intermediary causal mechanisms that might explain caveats patterns. Hence, we consider the inductive task of generalizing insights from the study of the historical record of particular cases, a necessary initial step in the development of research fields dealing with a recently acknowledged and scarcely studied phenomenon.

As indicated above, to generalize is to posit that idiographic explanations of particular cases may be nomothetically valid as well. To pull off this epistemological leapfrog, we need to probe the analytical usefulness of concepts to transcend the local experience. This implies creating "concepts at a sufficient level of abstraction to cut across local differences" (Platt 2007: 120). As mentioned, case-specific data also need to be inductively engaged in the theorizing about explanatory and contextual variables, causal mechanisms, interaction effects, and scope conditions (Collier 1999). In so doing, the case contests any fixed ties in time and space. "As a case of something," Jonathan W. Moses and Torbjørn L. Knutsen reason, the case "bows before theory and seeks to move from a purely empirical level of exposition to a level of general statements" (2012: 143). Subsequently, we test general statements against other cases belonging to the same class of phenomenon.

Inductive case-study reasoning does not make a claim to universal or statistical generalization. The nonexperimental and non-probabilistic nature of inference offered by single case-study design induces researchers to limit themselves to "narrow and well-specified contingent generalizations about a type of cases" based on logical inference (Blaikie 2009: 196).[3] A contingent or typological generalization specify

---

[3] J. Clyde Mitchell defines logical inference as the "the process by which the analyst draws conclusions about the essential linkage between two or more characteristics in terms of some systematic explanatory scheme – some set of theoretical propositions" (1983: 199–200). It follows from this that detailed knowledge of the context is a crucial element in the case study researcher's capacity to construct generic implications from a case.

the conditions which must be met for the generalization to be true, say the conditions that need to be fulfilled for states to apply a considerable amount of restrictive caveats. Case-study typological theorizing thus provides a means for specifying the configurations of variables to which generalizations apply (Hall 2007: 91–92). We use case-study findings incrementally to refine "middle range contingent generalizations, either by broadening or narrowing their scope or introducing new types and subtypes through the inclusion of additional variables" (George and Bennet 2005: 124).

In inductive reasoning, the specification of scope conditions requires serious consideration since scope conditions "constrain the applicability of general propositions by defining [limiting] the circumstances [boundaries] in which a theory is applicable" (Harris 1997: 123). Indeed, the initial problem of theory development is to determine the meaningful boundary of a case (Platt 2007: 116). As for theorizing about the politics of caveats, it is for instance not indifferent for theory development whether we decide to limit the study of caveats to multinational military enforcement operations only (as we have done), or to extend the concept to include also multinational peacekeeping operations. Note that military coalition operations are deployed to enforce a resolution on partisan terms. Multinational peacekeeping operations, on the other hand, are deployed to police a cease-fire, political elections, or a settlement already agreed upon by all legitimate parties to the conflict (Fermann 1992: 8). It is reasonable to assume that the conflict logic and the politics pervading such different kinds of coalitions are somewhat different. Hence, the broader conception of caveats would have required some additional theoretical reasoning as to what conditions might explain such an extended conception of caveats (George and Bennett 2005: 100–111).

The essence of generalizing from facts to theory is then to create an analytical language of concepts, typologies, and coherently linked ontological assumptions (theory) from the inductive study of a particular case (or cases), which is capable of describing and explaining a class of cases within the limits of some scope conditions. From the analytical constructs, empirical propositions (hypotheses) are subsequently deduced and tested against additional empirical material (other cases). What case-study designs may serve the task of generalizing beyond case-specific data in our particular field of research? We briefly consider three case-study techniques, which may serve the aim of theory development and hypotheses generation also in the study of the foreign politics of caveats.

Theory Building and the Deviant Case Design

The *deviant (or negative) case* procedure is a case-study design particularly useful for theory-developing purposes. The deviant case analysis include "studies of single cases that are known to deviate from established generalizations" (Lijphart 1975: 692). Data is analyzed through a process by which "theoretical ideas are revised and redesigned as data are compared with them" (Wicks 2012: 290). Following Jack S. Levy, deviant case-study designs focus on "observed empirical anomalies in existing theoretical propositions," to explain "why the case deviates from theoretical expectations and in the process refining the existing theory and generating additional hypotheses" (2008: 13).

Deviant case analysis allows us to show how data that run counter to theoretical explanations can be used to rethink explanatory variables and mechanisms, reconsider the operationalization of variables, distinguish between subtypes of the same phenomenon, and consider "whether and how the scope conditions of competing theories should be expanded or narrowed" (George and Bennett 2005: 115).[4] Indeed, the deviant case study can help us identify a new kind of case. For instance, the decision to include permissive caveats (as opposed to restrictive caveats) in the definition of national reservations on the use of force in coalition warfare was in part due to the fact that the United States was observed to apply more force or demand more control over the use of force than other members of the coalition. If we find this pattern of permissive caveats to be systematically confirmed by additional data/cases, the next question is if this deviant caveats pattern can be explained by the same contingent causal conditions as restrictive caveats patterns. Of course, the latter procedure is a theory-testing move.

However, the subsequent move would be inductive (theory-adjusting) in that the results from the study of the United States as deviant case would lead us to either celebrate the confirmed validity of the theoretical explanation, adjust the theory to better account for the deviating case of permissive caveats, or be forced to throw out permissive caveats from the definition of caveats, thus narrowing the scope conditions of the

---

[4] Such formulated it also becomes evident why we may use the deviant case procedure both for theory-developing and theory-testing purposes. Recall that juxtaposing theoretical constructs and facts is an epistemological two-way street (Moses and Knutsen 2012: 143), which also the plausibility probe and process tracing procedures discussed below are capable of traveling.

theoretical explanation. The latter two solutions qualify as theory developing contributions. Alexander L. George and Andrew Bennett drive home the main point of the argument: "Each case study [...] contributes to the cumulative refinement of the contingent generalizations on the conditions under which particular causal paths occur, and fills out the cells or types of a more comprehensive theory" (2005: 112).

George and Bennett also argue that the most general kind of finding from the deviant case study is the specification of a new concept, variable, or theory regarding a causal mechanism that "affects more than one type of case and possibly even all instances of a phenomenon" (2005: 114). When a deviant case study leads to the specification of a novel theoretical explanation, we may be able to generalize about how the newly identified mechanism may play out in different contexts. Coalition burden-sharing in the 1991 Gulf War serves well to illustrate the logic of the deviant case procedure for theory-developing purposes. Bennett et al. (1994) conducted a study of coalition burden-sharing in the 1991 Gulf War partly because several countries' sizeable contributions to the Desert Storm coalition contradicted the predictions of collective action theories that at the time dominated the literature on alliances (Olson 1971; Olson and Zeckhauser 1966). Collective action theory would have predicted more free riding. However, Bennett, Lepgold, and Unger found that pressure from the United States was the key to explain why states, such as Japan and Germany heavily dependent on the United States for their security, offered substantial contributions to the coalition war effort (1994).

While the temptation of free riding grows as one state becomes more powerful relative to others, so do the ability of the powerful state to coerce dependent allies as well. "As those forces offset one another, other factors, such as domestic politics and institutions, help tilt the balance towards or away from a contribution" (George and Bennett 2005: 115). In short, Bennett, Lepgold, and Unger developed contingent generalizations on how the understudied factor of alliance dependence would play out in different contexts, including the prospects of building domestic political support for a military contribution. The invitation to develop generalized contingent explanations of member states' coalition behavior is strong within the FPA approach precisely because the framework seeks explanations at several levels of analyses and in multiple research literature.

Theory Building and the Plausibility Probe

The *plausibility probe* was offered by Harry Eckstein (1975: 104–108) as a rather undemanding, exploratory theory-testing procedure to confirm that a theoretical hunch found some empirical support in a benevolent, "fitting" case (Moses and Knutsen 2012: 138–139). Turning the epistemological argument 180 degrees, we may also use the plausibility-probing reasoning as a vehicle for developing more solid theoretical generalizations, by systematically juxtaposing case facts and tentative analytical constructs. Indeed, Jack S. Levy argues that the *inductive* plausibility probe resembles a pilot study in experimental or survey research in that "it allows the researcher to sharpen a hypothesis or theory, or to refine the operationalization or measurement of key variables" (Levy 2008: 6). This is a crucial "intermediary step" before moving from hypotheses construction to time-consuming empirical tests (Eckstein 1975: 108). By probing "the details of a particular case in order to shed light on a broader theoretical argument" (Levy 2008: 6), inductive plausibility probes can be "cheap means of hedging against expensive wild-goose chases, when the costs of testing are likely to be very great" (Eckstein 1975: 110).

For instance, in developing a theory on national restrictions on the military use of force, we may use the inductive plausibility-probing case-study design to strengthen the theoretical arguments and sharpen the formulations of hypotheses prior to any comparative or large-N statistical study conducted for theory-testing purposes. Depending on the research question and the theoretical arguments made, Germany and Japan are excellent candidates for inductive plausibility probing and the deviant case-study design. This is because the Allies imposed strict limitations on the use of force in Germany and Japan's constitutions as part of the Second World War peace treaties. The institutional limitations on the use of force were to a considerable extent reflected in the complex political procedures necessary to make decisions on any support of (financial, logistical) or participation in multinational military operations, and in the subsequent inclination to apply restrictive caveats on troop contributions.

If the theory in question attempts to explain why states hesitate to join coalition forces, Germany and Japan would be perfect "laboratories" for the fine-tuning of theoretical arguments and sharpening of operational definitions within the logic of the inductive plausibility probe. If the theory aims at explaining multinational war-participating propensity in general, Germany and Japan would likely qualify as candidates for a deviant case study. The deviant case study would provide a check on the scope conditions for the theoretical argument. That is, how narrow or wide the empirical boundaries of the theory should be set.

Paying too little attention to plausibility-probing theory development prior to setting in motion large-N studies of foreign policy output may come at a considerable price. The main reason why FPA temporarily fell out of fashion in the early 1980s was the meager success of the Comparative Foreign Policy (CFP) project. CFP was one out of three main strands of FPA[5] and inspired by the work of James N. Rosenau on the relationship between genotypes of states and the sources of their foreign policy (1974). Valerie M. Hudson argues that the disappointing results from this empirical research program were the failure of the CFP project to develop a robust theoretical framework prior to the gathering and analyzing data (2005, 2007). On a general note, Charles C. Ragin argues that variable-focused researchers regularly refers to unobserved mechanisms on the case level to explain cross-unit patterns, and argues that if these mechanisms cannot be confirmed at case level, we should be suspicious about the across-unit empirical correlations arrived at by means of statistical analysis (2000: 28). Without prior case study of potential causal mechanisms linking independent and dependent variables, we risk spending a lot of time chasing data up blind alleys.

The hard lesson from the CFP experience is transferable to future research on the foreign politics of caveats, and possibly other conceptual frameworks that aim at explaining states coalition behavior and states' propensity to participate in multinational military operations. Indeed, two other main strands of the FPA approach invites detailed, prior empirical case-study research on the institutional and psychological mechanisms involved in foreign policy-making and implementing processes to understand better the causal paths and the mechanisms linking global and domestic input factors with foreign policy output and outcomes as regards caveats.

---

[5] In addition to the CFP strand of FPA, the other two strands include the study of institutional and psychological mechanisms involved in foreign policy decision-making (Smith et al. 2008: 3–4). More recently, efforts have been made to bridge the gap between IR and FPA (Thies and Breuning 2012; Rynning and Guzzini 2002; Rose 1998; Fearon 1998).

Theory Building and Process Tracing

Applied for theory-developing purposes, the *process-tracing case study* is geared toward the inductive "practice tracing" of real-life decision-making processes (Pouliot 2015). Here, the effort is to generalize case-specific empirical observations of "many mechanisms linked in causal processes" (Mjøset 2009: 58), develop "analytical narratives" to better understand why certain inputs are associated with certain outcomes (Bates et al. 1998), and use "evidence from within a case to develop hypotheses that might explain the case" (Bennett and Checkel 2015: 8).

Process tracing is thus a case-study technique aimed at "capturing causal mechanisms in action" (ibid.: 9). As such, process tracing may more than perhaps any other case-study procedure, compensate for the "soft underbelly" of rational choice approaches to the explanation of state behavior, and statistical large-N studies relying on this parsimonious approach to explain political behavior in general. In his seminal study of The Cuban Missile Crisis, Graham T. Allison (with Zelikow 1999) conducted inductive process tracing of the US decision-making and implementing processes to see what was going on between input and outcome in crisis management. Long before the case-study procedure of process tracing as inductive "practice" tracing was acknowledged by its present name, Allison shed light on decision-making processes to develop and later test diverging theories on foreign policy-making.

By shedding light on "causal-process observations" (Brady and Collier 2010: 12), we reach inside the "black box" of decision-making and explore governmental politics at work, the institutionalization of standard operating procedures (SOP) in decision-making, and the perceptions and expectations of actors. We may use process tracing to explain particular historical episodes, but crucially for our purposes also as a means to suggest generalizable causal hypotheses. Subsequently, we will return to the question of how the process-tracing case-study design be utilized for theory-testing purposes.

We have argued that the deviant case, the plausibility probe and the process-tracing case-study design be applied for theory developing as well as theory-testing purposes. Concerning theory development, the plausibility probe is not limited to a particular generalizing focus as it, in principle, covers any inductive reasoning from case-specific facts used to develop precise conceptual, typological, and ontological constructs for the subsequent study of a class of cases. Process tracing, on the other hand, is more specific in its theory-building implications. The case-study procedure of process tracing borrows from historicity, "the temporal structuration of social actions and processes" (Hall 2007: 82), in the attempt to conceptualize and theorize the social processes that causally connects independent and dependent variables.

By theorizing the timeline and causal path of case-particular decision-making processes, the process-tracing researcher would want to make sense of the connecting and mediating mechanisms at work, without which we cannot fully understand the presence of cross-unit empirical patterns. Understanding the causal chain of decision-making and implementing processes is at the core of FPA. As crucially, in the still-nascent field of research on the politics of caveats, theorizing, and testing of hypotheses on the mechanisms that may, or may not constitute the causal path from X to Y is insurance against spurious inferences in comparative case studies, and in statistical large-N studies further down the road.

Case Study for Theory Testing

The case-study practice of juxtaposing theory and evidence in an "alternating sequence of conjectures and refutations" (Levy 2007) implies that case studies play different epistemological functions at different stages of a research program. Having discussed how single case-study designs may be applied for theory-developing purposes, we are in a position to discuss how single and comparable-cases designs may facilitate the empirical testing of theoretical arguments and hypotheses. The aims of theory-testing procedures are to "strengthen or reduce support for a theory, narrow or extend the scope conditions of a theory, or determine which of two or more theories best explains a case, type or general phenomenon" (George and Bennett 2005: 109).

Common to single and comparable-cases designs (except process tracing), is that we assess causal inference by means of the *selective choice of cases* based on the nature of the research question, the theoretical argument and the hypothesis under consideration. In the test of a theory, it is essential to decide whether the test case in question qualifies as a "fitting" empirical plausibility probe or some considerably more demanding crucial case. Whether the troop-contributing nations of Germany or Norway are the right choice for this-or-that caveats case-study design are determined by the function the cases of Germany and Norway serve for the research question and the theoretical argument in question. Hence, the well-selected confrontation between theoretical argument and a particular case within the logic of a relevant case-study design is the key to proficient hypotheses testing in case-study research.

In contrast to the probabilistic logic of inference that underpins statistical large-N research, causal inference in single and comparable-cases designs rests on the logic of theory-guided pattern matching. In large-N research designs, which may represent only a fraction of the full population, we gather a representative sample of units through a random sampling procedure. In single and comparable-cases designs, we ask what case or cases would best illustrate, or put to the test our theoretical answer to the research question posed? This is precisely why case selection in qualitative research needs to be biased and very much dependent on the theoretical argument and the case-study design chosen. Causal inference relies on us selecting a particular case fitting both the theoretical argument and the case design well.

Theory Testing and the Plausibility Probe
In what capacity, then, may a case contribute to the refutation or strengthening of theoretical arguments and empirical propositions? A useful point of departure for an exposition on theory-testing case study is perhaps in the original application of Harry Eckstein's *plausibility probe* (1975: 104–108). In a previous section, we discussed how plausibility probes might contribute to theory building from the study of particular events (case as a subject). Having inductively established the generic constructs, the subsequent theory-testing angle to the plausibility probe is to look for a promising and novel case, which may serve as a preliminary and not-so-demanding trial-testing vehicle for the theoretical argument developed. In the plausibility probe, we seek out the theory-confirming case (Lijphart 1971: 691) to "investigate the degree to which a given case fits a general proposition" (Moses and Knutsen 2012: 137). This particular case-selection criterion induces us to look for a case that is likely to illustrate our theoretical arguments and initially confirm our hypotheses, thus reassuring us that we may be heading in the right direction analytically speaking. In turn, such assurance may justify more time-consuming investments in theory infirming (most- and least-likely case designs), deviant, process tracing, and comparable-cases studies (Levy 2008: 6).

Eckstein himself used the plausibility probe to seek anecdotal evidence for his "hunch" that democracy is more likely to thrive in societies that have "deeply-rooted egalitarian values, and less likely to evolve in societies that are marked by deep divisions and rigid hierarchies of authority" (Moses and Knutsen 2012: 138). His research-visit to Norway confirmed Norway to qualify as a "fitting" case for his theoretical argument and indicated that his evolving congruence theory was on the right track in the explaining of the conditions for political democracy. Acknowledging that Norway was a benevolent case for the testing of the congruence theory, Eckstein used the experience to refine his theoretical argument and added other, considerably more demanding cases to his testing regimen.

Closer to home, the plausibility-probing case study may be used to explore the validity of the proposition that coalition leaders are more likely to apply permissive caveats than junior members of coalition. The relevance of this hypothesis follows from the analytical decision to include permissive caveats into the broader conception of national reservations on the use of force. We may derive the empirical proposition from the argument that since the coalition leader must shoulder the lion-share of the military responsibility, it is likely to reserve itself some latitude to apply the additional use of force if circumstances so require. What would then be a "fitting" case to study if we were to test the permissive-caveats hypothesis within a plausibility-probing case design?

If the scope conditions of our theoretical argument were restricted to post-Cold War Western coalition-operations, the United States would be the obvious choice. If we were to find that the United States in, say Afghanistan, demonstrated a pattern of permissive caveats, we might be encouraged to pursue our research into much more demanding theory infirming (Lijphart 1971: 691), most-likely and least-likely theory-testing case-study designs. Subsequent to this, we may even decide to invest more heavily into a comparable-cases study of several nations' use of caveats across a handful of coalition operations while satisfying the requirement for variation on the dependent variable. If the deduced empirical proposition were to find support in the comparative analysis, this would strengthen our confidence in the theoretical argument considerably.

Everything else being equal, a comparable-cases design may provide a stronger test of a theoretical argument than single case designs. However, rarely is everything equal. For instance, a single case study of the United Nations Emergency Force II in Egypt (1973–1980) is better documented and deeper researched (Fermann 1988) than a comparative case study of the conditions for peacekeeping effectiveness that include

11 international peacekeeping operations (Fermann 1992). Fortunately, Harry Eckstein makes an argument for a theory-testing case design that is considerably more demanding on the theoretical argument than the theory-confirming plausibility probe, but which is considerably less costly to implement than a double-digit comparable-cases design.

Theory Testing, Crucial Case, Most-Likely and Least-Likely Designs (L3)
The *crucial case design* is a much more demanding test in that it is not content with seeking out a "fitting" or "theory-confirming" case. The crucial case design rather mimics the well-constructed, decisive experiment by selecting a well-chosen case, which can "provide strong support for, or falsify, a given theory" (Moses and Knutsen 2012: 139). How do we recognize a crucial case when we see it? In order to carry out a strategic test of a theoretical argument by means of crucial case design, Eckstein advises us to select a "case that *must closely fit* a theory if one is to have confidence in the theory's validity, or conversely, *must not fit* equally well with any rule contrary to that proposed" (1975: 118). Such formulated, crucial cases provides "the most definitive type of evidence" for the falsification or validation of a theoretical argument (George and Bennett 2005: 120), implying that it is "extremely difficult [...] to dismiss any finding contrary to the theory as simply deviant" (Eckstein 1975: 118).

However, truly crucial cases are hard to find in the social world. As a somewhat less demanding but more realistic alternative, Eckstein suggests studying *most-likely* and *least-likely cases*. George and Bennetts explain that "in a most-likely test, the independent variables posited by a theory are at values that strongly posit an outcome or posit an extreme outcome," while in "a least-likely case, the independent variables in a theory are the values that only weakly predict an outcome or predict a low-magnitude outcome" (2005: 121). Such cases still constitute tough tests capable of producing rather clear-cut results in terms of theory falsification and validation.

For theory-testing purposes, most- and least-likely case designs rest on the assumption that some cases are more important ("crucial") than others, and that "the weight of the evidence is relative to prior theoretical expectations" (Levy 2008: 12). The strongest possible supporting evidence for a theory in a tough test set-up is "a case that is least likely for that theory but most likely for all alternative theories, and one where the alternative theories collectively predict an outcome very different from that of the least-likely theory" (George and Bennett 2005: 121). More precisely, Jack S. Levy reasons that "if one's theoretical prior [expectation] suggest that a particular case is unlikely to be consistent with a theory's predictions, and if the data supports the theory, then the evidence from the case provides a great deal of leverage for increasing our confidence in the validity of the theory" (2008: 12). Conversely, "if one's prior [expectation] suggest that a case is likely to fit a theory, and if the data confound our expectations, that result can be quite damaging to the theory" (2008: 12). Hence, in the least-likely case design, the assumption and the inferential logic is that if a theoretical argument can make it there, it can make it anywhere. In the most-likely case design, the inferential logic is instead that if the theory tested cannot make it there, it cannot make it anywhere.[6]

The logic of inference in most/least-likely case analysis is not symmetric. The tougher test is the least-likely case because a confirmed theoretical prediction cannot also be ascribed to other, competing theories. George and Bennett argue that "theories which survive such a difficult test may prove to be generally applicable to many types of cases, as they have already proven their robustness in the presence of countervailing mechanisms" (2005: 121–122). As to the inferential value of the most-likely case, Levy reasons that "evidentiary support for a theory from a most-likely case (or lack of support for a least-likely case), [...] leads to only a modest shift in one's confidence in the validity of a theory" (2008: 12). However, more often than not cases "fall somewhere in between being most- and least-likely for particular theories, and so pose tests of an intermediate degree of difficulty" (George and Bennett 2005: 122).

George and Bennett argue that short of finding toughest and easiest test cases, "researchers should be careful to specify, for each alternative hypothesis, where the case at hand lies on the spectrum from most to least likely for that theory, and when the theory predicts outcomes that complement or contradict other theories' predictions" (2005: 122). Again, Graham Allison's study of US political decision-making during the 1962 Cuban Missile Crisis (with Zelikow 1999) shall serve to illustrate this pragmatic line of reasoning.

---

[6]Aptly, Jack S. Levy terms the inferential logic of the least-likely case the «Sinatra inference». As fittingly, the logic of the most-likely case is conceptualized the «inverse Sinatra inference» (2002: 442).

Due to the serious crisis-context of the foreign policy-making process, we may reasonably frame the Cuban Missile Crisis as a most-likely case for the rational actor model (RAM) of foreign policymaking. At the same time, political crisis management is a good candidate for a least-likely case for the competing organizational process model (OPM) and the bureaucratic politics model (BPM). George and Bennett argue that in these regards Allison's study is "a strong test of the rational actor model, a moderate test of the organizational process model, and a strong test of the bureaucratic politics model" (2005: 122).

As to the results of the study, "the evidence appeared to contradict many predictions of the rational actor model but to fit predictions of the organizational process and bureaucratic politics models, [which] increased scholar's confidence in the generalizability of the organizational and governmental politics models" (2008: 12). Both BPM and OPM came out strengthened from the analysis of a decision-making process characterized by unsurpassed high stakes, stress, and lack of time. The empirical tests confirmed several of these models' predictions in a situation much more likely to validate the predictions of RAM.

Had Allison selected a case of noncrisis decision-making, the testing of the predictions of BPM and OPM would not have constituted a "crucial" or "tough" case, quite the opposite. Routine decision-making in government agencies is the home turf of BPM and OPM. Uncovering evidence from a "fitting" case that is consistent with BPM and RAM, "would not have been surprising and consequently would not have significantly altered scholar's prior assessment of the broader validity of those models" (Levy 2008: 12–13). What was surprising was that RAM did not test very well on its home turf of crisis management, that BPM and OPM were to a considerable extent validated in a somewhat "misfitting" context, and that both models consequently came out increasing the boundaries of their scope conditions.

What would a promising candidate-case for a tough, least-likely test of the "fear of abandonment hypothesis" (AP-H1) deduced from Glenn H. Snyder's alliance dilemma argument that alliance members' inclination to contribute to the common cause of collective security is a function of the particular balance between fear of abandonment and fear of entrapment (1984) look like? Recall that the AP-H1 hypothesis asserts that, "states primarily motivated by strong fears of abandonment are inclined to provide unconditional military support to the coalition (no caveats)." An alternative, competing hypothesis based on the assumption that "all politics is local" (Tipp O'Neill) might read as follows: "States with constitutional and procedural restraints on the participation in coalition forces are (also) more prone to apply restrictive caveats when participating in coalition operations." We may dub this proposition "der Primat der Innenpolitik" hypothesis, following Eckardt Kehr's finding that domestic industrial interests had a decisive influence on policies directing the German naval build-up in the 1890s (2012 [1927]; Lombardi 2008). To construct a tough, least-likely test of the outside-in "abandonment" hypothesis, we need to identify a case which is unlikely to confirm the hypothesis, and simultaneously likely to confirm the alternative inside-out "der Primat der Innenpolitik" hypothesis.

Unambiguous tough test cases are hard to find, but let us for the argument assume that Germany due to its geographical position in Central Europe and its indispensability as a NATO member can allow itself a considerable degree of free-riding behavior in alliance politics. We would thus expect Germany to apply caveats, if participating with any substantial forces at all, much more often than most other NATO members. At the same time, Germany would even more convincingly qualify as a most-likely case for the competing "der Primat der Innenpolitik" hypothesis due to the relatively strong constitutional and procedural limitations on the participation and use of force in coalition out-of-area operations. We would expect that Germany, for historical and domestic political reasons, participate with one arm tied to its back. That is to say, apply restrictive caveats, if contributing to the coalition at all.

If it is actually found that Germany in the imagined coalition force not only decides to contribute a substantial military contingent to the multilateral effort, but also refrain from applying any restrictive caveats on the use of force, the case of Germany would constitute a strong empirical test of the "abandonment" hypotheses, and indeed *validate* this empirical proposition. What makes the test a tough test is that the case of Germany simultaneously would *negate* the alternative explanation related to the primacy of domestic politics, the "der Primat der Innenpolitik" hypothesis.

Our discussion on most- and least-likely case-study research design suggests that a small number of case studies, and possibly even a single case, "can be quite valuable [in the testing of] certain types of theoretical propositions" (Levy 2008: 13). Indeed, John Gerring argues that most/least-likely cases provide "the strongest sort of evidence possible in nonexperimental, single-case study" (2007: 115). What more can *comparable*-cases designs offer in terms inferential power in the empirical testing of theory?

Theory Testing, Comparable-Cases Designs, and Qualitative Comparative Analysis

All science, Political Science included, is about comparing cases and about juxtaposing analytical constructs to empirical patterns. As noted by Moses and Knutsen, classical examples are Aristotle (1979 [350 BCE]), who compared a large number of constitutions in the ancient world to identify the best and most stable types, and Machiavelli (1961 [1532]) who compared rulers' behavior to identify the governing principles for efficient rule (2012: 95). John P. Frendreis puts comparative methodologies in context by arguing that the comparative method is a strategy for imposing scientific control "for potential confounding variables through careful selection and matching of cases rather than through experimental manipulation or partial correlations" (1983: 255). Moses and Knutsen remind us that the comparative method nevertheless is an attempt to "mimic the scientific logic of experimentation, but without the ability to fully control the test-situation" (2012: 116).

Whereas control for third variables in the laboratory experiment is established by erecting glass walls (proverbial or real) around the substances included in the experiment, in comparable-cases design we achieve control by way of case selection. Still, John Stuart Mill acknowledges the epistemological kinship between the experiment and the comparative theory-testing logic by naming the relevant chapter in his study *A System of Logic* "The Four Experimental Methods" (2002 [1891]). Here Mill elaborates on different designs for the comparative control for third variables. As to the distinction between statistical and comparative theory-testing methods, the former relies on the random selection of a representative sample of cases and probabilistic relationships, while comparable-cases designs rely on a pattern-matching logic of causal inference that requires the careful selection of cases so as to "maximize the variance of the dependent variables and to minimize the variance of the control variables" (Lijphart 1975: 164).

The main pitfalls of comparative analysis are the sampling bias in the selection of cases resulting from the relaxing of comparability criteria, and the overdetermination problem arising when the comparative analysis depends on too few cases relative to the number of explanatory variables. Sampling bias can threaten the generalizability of any results we might produce. The solution to the problem of sampling bias is to refrain from lowering the threshold for what qualifies as comparable cases too much. The solution to the overdetermination problem is to increase the degrees of freedom, either by increasing the number of cases by means of diachronic or synchronic extension, or by reducing the number of explanatory variables included in the comparative analysis (Collier 1993: 111–113). The latter we achieve by fusing variables that in part seems to address the same phenomenon, or by omitting variables that do not enjoy the support of a strong theoretical argument (Lijphart 1971: 685).

Comparable-cases research designs come in the basic shapes of the Method of Difference (MD), the Method of Agreement (MA), the Joint Method of Agreement and Difference (JMAD), and the Method of Concomitant Variation (MCV), which were reasoned by John Stuart Mill (2002 [1891]) and to varying degrees iterated and elaborated on in the "different systems/similar systems" approaches (Przeworski and Teune 1970); the "comparable-cases strategies" (Lijphart 1975); the "focused comparisons" (Hague et al. 1998: 280); the case-oriented comparisons (Ragin 1987); and the "method of systematic comparative illustration" (Smelser 1973).[7] More recently, QCA has been developed to

---

[7] Note that in the subsequent discussion, we emphasize comparative methods as means of testing theoretical arguments and empirical propositions that can explain patterns and regularities in the social world. Central to such a *realist* (or naturalist) epistemological position is that we are interested in comparative methods in relation to how case selection and choice of data can contribute to the validation of causal explanations in a nomothetic sense. This is in considerable contrast to the *constructivist* approach to comparative methods, where the focus is not so much on comparative control for third variables "as for how to preserve and exploit the qualities associated with thickly descriptive narrative" (Moses and Knutsen 2012: 231). The constructivist approach to comparative methods would include emphasizing the "uniqueness, particularity and complexity of social and political phenomena," and how comparisons can be used "in a less formal sense to challenge existing explanations and to explore possibilities" (ibid.: 245). Finally, constructivist approaches to the social and political world also engage in how meanings are embodied in agency rather than taking material facts at their face value. Hence, constructivist comparative methodology rest on a hermeneutic epistemology, draws on narratives and representations of meaning, and is closer to the idiographic research ambition than the nomothetic. We acknowledge that the constructivist approach to comparative methodology is certain to offer insights as to how we usefully can research the politics of caveats. However, while the epistemological foundation of the empirical research program does not a priori exclude any middle-ground constructivist approach capable of being integrated into an empirical research program (Adler 1997), we dedicate the remainder of the methodology discussion to the uncovering of causal relationships between variables rather than to how the construction of meaning be empirically researched.

deal with a larger number of cases short of statistical large-N studies, and to better cope with causal complexity.

Comparable-cases designs thus come in different shapes, but unite in the ambition to widen the empirical basis for making causal inferences while simultaneously preserving some of the empirical depth and contextual richness characterizing single and few cases designs. Contextual richness and the capacity to unravel different causal paths that leads to the same outcome are what makes comparable-cases designs not merely an intermediary step toward the statistical large-N study, but a design with its own particular epistemological advantages.

Recall that John Stuart Mill offers four main approaches to the comparative control for third variables: The Method of Difference (MD), the Method of Agreement (MA), the Joint Method of Agreement and Difference (JMAD), and the Method of Concomitant Variation (MCV). Arguably, the first method is less applicable for research on social phenomena. Only the fourth method is capable of coping with data on the interval measurement level. As for the former three comparative techniques, they are limited to research on binary cases; that is, limited to the processing of data on the nominal level of measurement only. This implies that the former three comparative techniques cannot track variation in magnitude but merely in the simple presence or absence of a variable (Moses and Knutsen 2012: 97–102).

Theory Testing and the Method of Difference
Scholars have used the MD in diachronic/longitudinal comparisons of conditions of a single country in two or more different points in time; to compare intrastate differences; to compare polities across nations that are relatively similar; and in counterfactual analysis where a real case is compared to an imagined case that is exactly similar. In the MD, we select cases on the condition that they have different values on the dependent variable and similar values on all but one of the possible causal variables (Mill 2002 [1891]: 256). To the extent that these ideal-typical selection conditions be fulfilled, the MD technique eliminates extraneous variables that vary across cases. The basic inferential logic of the MD is to "identify patterns of covariation and to eliminate independent variables that do not co-vary with the dependent variable" (Levy 2008: 10). To identify patterns of covariation, we need to look for the decisive, remaining difference between the compared cases which can explain the variation in the dependent variable.

Mill himself was skeptical about applying the MD in the study of social phenomena because he considered it nothing less than "manifestly absurd" that cases at the aggregate level of nations could fulfill the demanding criterion that they are similar "on all but one of the possible causal variables" (Mill 2002 [1891]: 575). Nevertheless, scholars have used the matching strategy of the MD extensively in comparative research, in particular within Area Studies where similarity across cases can be argued with some credibility.

Within our research group, Gabriel Husby (2015) conducted a comparative case study of Norwegian caveats behavior in four coalition operations in Kosovo, Afghanistan, and Libya. Arguably, the comparative study fits the case selection criteria of the MD reasonably well. The diachronic comparative study reveals that the Norwegian government from 1999 (Kosovo) through 2011 (Libya) attached increasingly less restrictive caveats on her contribution to Western coalitions, especially as regards the air force detachment. From only "moving air" over the Adriatic in a recumbent role in 1999, Norwegian F-16 fighter jets dropped some 580 bombs on Libya twelve years later. In between the missions in Kosovo and Libya, The Norwegian Air Force dropped seven bombs during the several years of deployment in Afghanistan.

How can this gradual and increasingly dramatic rise in offensive military activities be explained? In his comparative analysis, Husby finds strong indications that the development from "minimal" bombing in Afghanistan to "maximum" military impact in the Libyan theater of war mainly can be explained by changes in the balancing of the alliance dilemma (Snyder 1984) as perceived in Oslo. The Norwegian Government's concerns about "abandonment" were increasing relative to its concerns about being "entrapped" in a coalition including several important and close NATO allies. There was a perception among key policy-makers that Norway needed to prove herself in the burden-sharing alliance context. Hence, the use of restrictive caveats on Norwegian air operations grew increasingly irrelevant, if not politically counterproductive. As to the "absence" of offensive use of the Norwegian Air Force in Kosovo, we cannot easily understand this pattern in terms of caveats. This is because the Norwegians at this time lacked the laser-guided bombing technology and night-flying capabilities to deliver offensive action without severe risk of collateral damage and loss of aircraft, especially given the strong Serb air-defenses on the ground (Husby 2015: 145–152).

Another interesting study is Tony Ingesson's research on the rein-
forced Swedish–Danish–Norwegian mechanized battalion (Nordbat)
deployed in Bosnia as part of the ongoing UN peacekeeping mission
UNPROFOR (United Nations Protection Force). The study is designed
as an idiographic single-case study of "one of the most trigger-happy
UN units in Bosnia" (2016). The aim is to explain what, from our
perspective, looks like a pattern of permissive caveats, or at least excessive use of force relative to the coalition RoE. Ingesson finds that the
"trigger-happiness" of the Swedish-led and dominated Nordic battalion is conditioned by a "well-entrenched culture of mission command"
that emphasizes offensive action and delegates a considerable degree of
autonomy to the military commander (a Swede) to make operational
decisions on the spot (2017).

This culture of mission command was cultivated and institutionalized
in a Cold War environment where Swedish forces were expected to continue fighting against Soviet forces even if the lines of communication
to headquarter were cut, and also if rumors of capitulation were substituting access to solid information. In the complicated context of Bosnia,
this energetic military doctrine translated into a Nordic battalion that
confronted Serb forces when the Swedish battalion commander considered it necessary, thus deviating from the considerably more evasive military posture practiced in other contingents in UNPROFOR.

How does this single case study relate to our discussion on the nature
and application of the MD comparable-cases technique? Arguably, the
single case-study of Nordbat would have benefitted from being redesigned into a systematic MD comparable-cases study, thus increasing
comparative control. By making each contingent of Nordbat into a separate case (diachronic extension), we might systematically track changes
in the dependent variable (use of force) against any decisive changes in
explanatory variables.[8] One explanatory variable is of particular interest: That is continuity and change in the military command of Nordbat.
Ingesson can document that it took an extraordinary political intervention to substitute a row of battalion commanders cultivated in the energetic Swedish tradition of command with commanders more receptive to
political constraints on the use of force.

---

[8]Everything else being equal, a diachronic extension of cases increases the inferential
power of the study and increases the degrees of freedom. However, a diachronic extension
of cases also introduces the problem of non-independent observations (autocorrelation).

As to the question of national reservations on the use of force, it is problematic to conceive of the offensive posture of the earlier generations of Nordbat as a pattern of permissive caveats. This is because the offensive posture was most likely causally related to Swedish military culture and doctrine, and not traceable to a calculated political decision. Neither does it seem reasonable to interpret the more defensive military posture of the later generations of Nordbat as a pattern of restrictive caveats. Not because of any lack of political intent, but because Nordbat's use of force at this time did not deviate much from the coalition ROE. The rather stand-offish coalition RoE was already reflecting the very delicate political situation in the Balkans.

Theory Testing and the Method of Agreement
The basic inferential logic of the two comparable-cases designs of the MD and the MA is the same, to identify patterns of covariation and to eliminate independent variables that do not covary with the dependent variable. However, while we in the MD use similarities across contexts to find the one differing variable to explain an empirical pattern, in research employing the MA we use the "many differences found across cases to isolate a [decisive] common feature, the one variable that co-varies" with the pattern of caveats "across each of the otherwise disparate cases" (Moses and Knutsen 2012: 103). In the comparative-cases design of the MA patterns of covariation are then revealed by focusing on "cases that are similar on the dependent variable and different on all but one of the independent variables," thus eliminating "extraneous variables that do not vary across cases" (Levy 2008: 10). By applying case-selection criteria emphasizing some *decisive* difference or similarity in covariation across cases, both the comparable-cases designs of the MD and the MA invite dichotomous variables, produce binary cases, and lead to the subsequent loss of nuanced information.

Moses and Knutsen argue that the MA is not "encumbered with the same sorts of strict conditions as we saw with the MD," thus "lending itself more easily to Social Science." However, the MA is generally regarded as inferentially "inferior to the MD because it has a tendency to lead to faulty empirical generalizations" due to its incapacity to establish "any necessary link between cause and effect" (2012: 102–103). In particular, the MA is incapable of dealing with the challenge of multiple causation (or equifinality)

(Ragin 1987). For such reasons it can be argued the MA should be regarded primarily as a method for the *elimination* of explanatory candidates.

Still, the MA has been applied on several occasions, for instance in explanatory research on political revolution (Brinton 1965; Wolf 1968). More frequently, we have seen the MA applied in comparative research in *combination* with other methods for control of third variables. Our study of the conditions for effective international peacekeeping may illustrate how (Fermann 1992: 121–130, 178–198). First, we reasoned that the number of cases could be extended diachronically from 11 to 13 because the United Nations Emergency Force I (UNEF I) and the United Nations Force in Cyprus (UNFICYP) underwent deep changes due to force majeure, and thus could justifiably be counted as two each. Second, we applied both MD and MA techniques to maximize the number of pairwise combinations of cases that fulfilled said comparability/matching criteria to strengthen the inferential validity of the study. Third, applying a comparative technique resembling a crisp-set QCA[9] (Rihoux and De Meur 2009), we constructed a truth table to sort cases by the combinations of causal conditions they exhibit. Finally, the pattern-matching configurations were computed, and covariation expressed as gamma coefficients (-1 through 1) for each one of the six relationships hypothesized.

George and Bennett argue (and recommend) that qualitative research usually involves a combination of cross-case comparisons and within-case analysis "using the methods of congruence testing [diachronic comparison] and process tracing" (2005: xiv). The study referred to above shows how research combining synchronic and diachronic comparable-cases techniques may be designed. In the final section of the chapter, we will indicate how more advanced comparable-cases designs (QCA) may benefit from the high resolving power of the process-tracing technique. Prior to this, John Stuart Mill has more to offer on comparable-cases design.

Theory Testing and the Joint Method of Agreement and Difference
Mills' third comparable-cases selection method is the *Joint Method of Agreement and Difference* (JMAD). The JMAD relies on a mirror

---

[9]A crisp-set of data implies that data is dichotomized and thus that data is at the nominal level of measurement.

application of the Method of Agreement (MA) elaborated on in the previous. Applying the comparable-cases procedure of the JMAD, we can systematically compare all cases for both agreements as well as differences,[10] thus allowing for the inclusion of negative cases. Indeed, Moses and Knutsen argue that "the major difference between the JMAD and the [considerably weaker] MA is that the Joint Method uses negative cases to reinforce conclusions drawn from positive one" (2012: 107). Mill himself describes the procedure in considerable detail:

> If two or more instances in which the phenomenon occurs have only one circumstance in common, while two or more instances in which it does not occur have nothing in common save the absence of that circumstance, the circumstance in which alone the two sets of instances differ is the effect, or the cause, or an indispensable part of the cause, of the phenomenon. (Mill 2002 [1891]: 259)

If it by way of the MA method is found that the presence of X is the likely cause of Y, and it also can be shown through the MD procedure that in cases where X is not present neither is Y, we have conducted a very strong comparative test of the theory in question. Indeed, by "juxtaposing positive and negative cases [...] we can be more certain of the causal relationship at work" (Moses and Knutsen 2012: 107).

The JMAD procedure has been used in comparable-cases selection design in the acclaimed works of Barrington Moore and Theda Skocpol. Moore used the design in his study on the social conditions for dictatorship and democracy (1966), as did Skocpol in her study of social revolutions and their origin in great-power war, peasant rebellion, and rifts between elite groups, and subsequent state collapse (1979).

As for research on coalition politics, recall our study on conditions for effective international peacekeeping (Fermann 1992), which has been referred to previously to illustrate several points: To argue the importance of clarifying the scope conditions of theories, explain the advantages of comparative research, and to show how Mill's MA can be applied in conjunction with other case selection and statistical techniques where required or possible. We shall make a final point on the back of the comparative peacekeeping study. By utilizing both the MD and the

---

[10]The Joint Method of Agreement and Difference (JMAD) has also been referred to by Mill as the Indirect Method of Difference (IMD).

MA as case-selection criteria (pairwise comparability criteria), the comparable-cases study also satisfies the basic conditions for a JMAD comparable-cases design.

Theory Testing and the Method of Concomitant Variation
Moses and Knutsen observe that Mill's *Method of Concomitant Variation* (MCV) is "remarkably similar to the statistical approach" in that his final comparable-cases design is capable of observing "the quantitative variations of the operative variables" (2012: 109, 110). This implies that the MCV is not limited to binary cases, but can track variation in magnitude rather than in the simple presence or absence of a trait, which is a serious limitation in the comparable-cases selection techniques of the MD, MA, and the IMD discussed in the previous. Mill explains the MCV procedure as follows:

> Whatever phenomenon varies in any manner whenever another phenomenon varies in some particular manner, is either a cause or an effect of that phenomenon or is connected with it through some fact of causation. (Mills 2002 [1891]: 263)

This formulation points in the direction of continuous variables and degrees of causal relationships, and consequently requires data on the interval measurement level. Moses and Knutsen argue that while Emile Durkheim was "skeptical about the applications of the Method of Agreement and the Method of Difference to social phenomena, his skepticism did not extend to Mill's Method of Concomitant Variation" (2012: 109). Gabriel Almond and Sidney Verba used the MCV in an *inductive* way to come to grips with the underlying commonalities related to the political culture associated with democracy (1963). In *The Civic Culture*, they "followed J.S. Mill's lead [...] by operationalizing a number of very slippery and amorphous concepts, such as 'pride'" (Moses and Knutsen 2012: 111). The clarification of concepts was a precondition for the subsequent surveying of compatible data on political attitudes in five countries. Subsequently, the descriptive statistical data was put to work and showed that "democracy is most stable in societies where participation is tempered by [a 'civic culture' characterized by] elements of subject and parochial attitudes" (ibid.).

The MCV method observes and measures the quantitative variations of the variables. Hence, the most advanced of Mill's comparable-cases techniques is very close to resembling a statistical method, save the sufficient number of units. The development of modern statistics came after Mill and provided a much more potent toolbox for identifying the factors Mill was seeking to identify. Still, Mill's basic matching-strategies are useful to clarify our understanding of causal relationships, and how to test for covariation by means of comparable-cases design.

Shortcomings in Mill's arsenal of comparable-cases techniques are nevertheless evident at this point. First, a major problem confronting any comparable-cases research design is the difficulty of identifying cases that are truly comparable—identical or different in all respects but one. Second, the case-selection rules presuppose that we have a list of candidate cases to consider. However, the rules themselves do not tell us how to come up with such a list. If it were not for our previous theory-building (inductive) single-case studies, we would have no well-developed theory to deduce empirical propositions from saving the hunch of an informed guess. Third, Mill's comparable-cases techniques assume that among the list of factors under consideration, only one factor on the list is the unique cause of the effect. This is an unrealistic assumption in a Social Science with few if any law-like relationships. The more interesting research questions we come up with tend to deal with complicated puzzles involving several factors. Such puzzles are more likely than not to find its answer in the causal plural.

Following from our first and third point, we finally have no guarantee that the single relationship tested for is a causal relationship even if confirmed in the empirical analysis because it cannot be ruled out that covariation is due to some other factor hidden in an invalid assumption of comparability (spuriousness). For such reasons, several of Mill's matching strategies are better suited for the *elimination* of causal candidates than with confidence supporting causal relationships. We are thus inclined to consider Mill's methods probative, and not definitive as regards theory testing.

This leads to a methodological crossroads. If we have a large enough number of units/cases at our disposal, we can exploit statistical techniques for control of third variables in research. However, if our field of study does not yet possess systematic and reliable data to the extent necessary to run statistical analyses, we are left with QCA to move research on the (foreign) politics of caveats forward. Preferably in combination with other case designs.

Theory Testing by Way of Qualitative Comparative Analysis (QCA)
We should not consider QCA an inferior choice to statistical control. QCA can deal with research questions engaging too many cases (double-digit) for researchers to keep all the case knowledge in their heads, but too few cases for conventional statistical techniques. Following Ragin (1987, 1994, 2007, 2008), Byrne and Ragin (2013), and Rihoux and Ragin (2009), QCA is a bridge between qualitative and quantitative methods, and based on the analysis of set relations, not correlations. QCA requires familiarity with cases and some in-depth knowledge. Simultaneously, QCA is capable of detecting decisive cross-case patterns, the usual domain of quantitative analysis. In the examination of cross-case patterns, QCA accounts for the diversity of cases and their heterogeneity with regard to their different causally relevant conditions and contexts by comparing cases as configurations (combinations) of values on several variables (Ragin 2000: 67–82; Elman 2005).

Note that important causal relations, necessity, and sufficiency, show when certain set relations exist: With necessity, "the outcome is a subset of the causal condition." As for sufficiency, the "causal condition is a subset of the outcome" (Ragin 2000: 22). Due to its Boolean-based and pattern-matching logic (Ragin 2000), QCA is better than the probabilistic logic of inference that underpins statistical methods dealing with causal complexity in what Mill pointed out as "plurality of causes" (2002 [1891]), and in general systems theory is known as "equifinality" (Lieberson 1992). Ragin himself (1987) terms such causal complexity as "multiple conjectural causation." He argues that standard statistical methods cannot easily deal with this phenomenon because the number of interaction terms necessary to capture combinatorial effects increases rapidly with the number of variables and thus quickly overwhelms the degrees of freedom.

Ragin developed "qualitative comparative analysis" based on Boolean algebra to identify and test combinatorial hypotheses. QCA allows for pattern-matching strategies that deal with complex causation involving interaction effects, and "several different sets of causal conditions that may lead to the same outcome" (George and Bennett 2005: 162–163). We may use QCA to uncover different sets of conditions (INUS conditions) that explain patterns of caveats.[11] Such sets of conditions are

---

[11] An INUS condition is a causal condition that is "an insufficient but necessary part of a condition which is itself unnecessary but sufficient for the result" (Mackie 1965: 245).

"different configurations of factors conditioning patterns of [caveats] through different causal paths" (Platt 2007: 115).

Arguably, explanatory complexity is the rule rather than the exception in the study of social and political phenomena, foreign policy decision-making included (Haney 2002). Indeed, Jack S. Levy argues that claims of conditional necessity and sufficiency are common in theories involving causal complexity in which there are "multiple paths to an outcome and the presence of one condition might be a necessary condition for the impact of another variable along that particular causal path" (2008: 9). The multi-level and decision-making approach of FPA directs us to include and prepare for such causal complexity in the subsequent choice of several bodies of theory and research methods. Recall that research on the foreign politics of caveats currently also lack the data required to run statistical analyses of caveats-relevant relationships. Hence, in this regard, QCA and a research program grounded in the FPA approach would seem to be an extremely good fit.

The QCA methodology excels in the 5–75 cases range and when complex causation is theoretically anticipated. The strategy is to unravel configurational causation by establishing set relations. QCA can handle data on the nominal, ordinal, and interval level of measurement. This corresponds to crisp-set, multi value-set, and fuzzy-set QCA. While the technicalities differ, the basic procedure of the three QCA techniques is quite similar. Following Ragin closely (undated), the first step is to identify relevant positive and negative cases on the outcome, and the various combinations of conditions, the "causal recipes," that might generate the outcome.

Second, we construct a truth table based on the causal conditions specified or some reasonable subset of these conditions. A truth table sorts cases by all the logically possible combinations of causal conditions they exhibit, including those without empirical instances. We then assess the consistency of the cases in each row with respect to outcome. To what extent do they agree? We identify contradictory rows, and compare cases within each contradictory rows to identify decisive differences between positive and negative cases, and revise the truth table accordingly (Ragin, undated).

Third, we analyze the truth table. The goal is to specify the different combinations of conditions linked to the selected outcome, based on the features of the positive cases that consistently distinguish them from the negative cases. Paired comparisons continue until no further simplification is possible. We eliminate the paired comparisons that differ. The process of paired comparisons culminates in a list of causal combinations linked to the outcome. The truth table is thus the key tool for systematic analysis of causal complexity. Truth tables "list the logically possible combinations of causal conditions," such as the presence or absence of some threat perception (AP-H; AP-H3), decision-making autonomy (DGP-H3), or willingness to accept risk (PI-H6)), "along with the outcome exhibited by the cases conforming to each combination of causal conditions." That is to say, whether the application of national reservations on the use of force is "consistently present among the cases displaying each combination of conditions" (Ragin 2011: 73).

The final step is to evaluate the results of the previous elimination process. We interpret the results as causal recipes. That is *combinations of conditions* that might generate the outcome. Do the combinations make sense to us? What causal mechanisms do they entail or imply? How well do they relate to existing theory? Do they challenge or refine existing theory? Then we identify the cases that conform to each causal recipe. Do the recipes group cases in a meaningful way? Do the groupings reveal aspects of cases that have not been considered before? " *We conduct additional case-level analysis with an eye toward the mechanisms implied in each recipe. However, causal processes we can study only at the case level, and should be evaluated at that level*" (our Italics) (Ragin, undated).[12]

Theory Testing and Process Tracing

The reason we have postponed the discussion of process tracing as theory-testing case design becomes apparent in the two preceding sentences. Process tracing excels and supplements where both QCA and statistical analyses show their limitation: At the case level, process tracing is capable of "capturing causal mechanisms in action" were comparative and statistical techniques cannot (Bennett and Checkel 2015: 9).

Ragin argues that scholars focusing on variables, as a matter of routine assumes theoretical mechanisms to explain empirical patterns across cases (2000: 28). However, if we are not able to confirm such mechanisms on the case level, the results from statistical and comparative analyses cannot be entirely trusted (Møller 2015: 127–132). The merit of

---

[12]For detailed expositions on QCA crisp-set, multi-value set, and fuzzy-set procedures confer Rihoux and De Meur (2009), Cronqvist and Berg-Schlosser (2009), and Ragin (2009).

the within-case method of process tracing is its capacity to rule out and eliminate spurious inferences not traceable by means of comparative and statistical control for third variables. How is this particular case design capable of tracing causal paths between input and outcome? Moreover, how may process tracing contribute to research on caveats? Finally, to what extent does the case design of process tracing provide an answer to the directional requests of the FPA approach that we take the study of foreign policy decision-making and implementing processes seriously?

Andrew Bennett and Jeffrey Checkel define process tracing as "the analysis of evidence on process, sequences, and conjectures of events within a case for the purposes of either developing or testing hypotheses" (2015: 7). Alexander L. George (1979: 46) developed process tracing as a method for assessing the mechanisms used to argue claims about causation between two phenomena (variables). George considers a causal relationship something more than a covariation between two variables controlled for a number of third variables. Rather, he understands causation in terms of a *causal path* made up of a shorter or longer string of mechanisms connecting independent and dependent variables (Brady et al. 2010: 15–32; George and Bennett 2005: 214–215).

By adding process tracing to our research tool-box, we engage in the uncovering of "linear causal processes embedded in time" (Moses and Knutsen 2012: 134), where "A causes B, B then causes C, C then causes D and so on" (Checkel 2006: 363). By so doing, process tracing shifts our epistemological focus from *what* happened, to *how* and *why* it happened (Moses and Knutsen 2012: 225). The theory-testing purpose of process tracing is indeed to examine "the observable implications of hypothesized causal mechanisms within a case to test whether a theory on these mechanisms explains the case" (Bennett and Checkel 2015: 7).[13] By so doing, the process-tracing technique at the level of the single case makes us more convinced that statistical covariations and comparative pattern-matching inferences are causally connected and how.

---

[13] Recall that there is a theory building side to process tracing as well. Applied for theory-developing purposes, the process-tracing case study is geared toward the inductive "practice tracing" of real-life decision-making processes (Pouliot 2015). Here, the effort is to generalize case-specific empirical observations of "many mechanisms linked in causal processes" (Mjøset 2009: 58), develop "analytical narratives" to better understand why certain inputs are associated with certain outputs (Bates et al. 1998), and use "evidence from within a case to develop hypotheses that might explain the case" (Bennett and Checkel 2015: 8).

Levy argues that the process-tracing case design has a comparative advantage in the empirical analysis of "decision making at the individual, small group, and organizational levels," including the analysis of leaders' perceptions, judgments, preferences, internal decision-making environment, and choices (2008: 11). Pascal Vennesson concurs in the assessment that process tracing "provides a way to learn and to evaluate empirically the preferences and perceptions of actors, their purposes, their goals, their values and their specification of the situations that face them" (2008: 233). For such very process-tracing purposes, we may use "histories, archival documents, interview transcripts, and other sources to see whether the causal process a theory hypothesizes or implies in a case is in fact evident in the sequence and values of the intervening variables in that case" (George and Bennett 2005: 6). As process-tracing evidence, we would include a variety of within-case evidence of a temporal, spatial, or topical nature (Bennett and Checkel 2015: 7–8).

As indicated, we may combine process-tracing case designs with QCA comparable-cases design to examine empirically the alternative causal mechanisms associated with observed patterns of covariation. It is also effective in discovering and explaining deviant cases regarding causal paths and outcomes. Attempts to combine large-N statistical studies with case studies involve process tracing (Sambanis 2004). Similarly, attempts to combine formal modeling approaches with case studies also utilize process tracing, in part to validate the preferences and decision-making calculus attributed to political leaders and other actors (Bates et al. 1998; Brams 1994; de Mesquita 2000).

As elaborated on above, process-tracing case design is a door opener to the proverbial "black box" of governmental decision-making and implementing processes. As QCA responds to the multi-level explanatory ambition of FPA by allowing for "multiple conjectural causation" (Ragin 1987), process tracing is tailor-made to deliver on this the second main tenet of the FPA approach that underpins our empirical research program. In particular, process tracing delivers by tracing the political, institutional, and psychological mechanisms at work in the molding and translation of external and domestic input impulses to the foreign policy decision-making and implementing processes that eventually produces some coalition-behavior outcome and some caveats pattern. On the road to comparative and statistical research, we should also consider mining the potential in applying process tracing inductively as "practice tracing" to feed Alexander L. George's method of "structured, focused comparison" with relevant policy-making and implementing data (George and Bennett 2005: 67–72). "Structured, focused comparison" is tailor-made for theory building in the study of foreign policy-making processes, and thus a fruitful point of departure for generation of hypotheses on the foreign politics of caveats.

Per Marius Frost-Nielsen used a combined structured focused comparison, and process-tracing design for theory testing on cases not already contributing to theory building. In a recent comparative study of the application of caveats on Denmark's, the Netherland's, and Norway's contributions to NATO's intervention in Libya in 2011, he finds that "caveats are linked to three different causal pathways" (2017: 3) in that "domestic factors help to explain whether or not there will be caveats, while external pressures help to explain the form that such caveats take" (ibid.: 22). As to the politics of implementation, Frost-Nielsen concludes that "in lack of clear guidance from their national principals, military officers might themselves apply reservations in anticipation of what their political authorities find acceptable" (ibid.). At the more overriding level, his study also confirms that the phenomenon of caveats is conditioned through "multiple conjectural causation," and consequently that the multi-level and decision-making approach of FPA is a fruitful analytical point of departure to study the politics of caveats. Indeed, Frost-Nielsen suggests that "caveats are best accounted for through a typological theory by distinguishing between several types of caveats based on different causes or configurations of causes that led to caveats" (2017: 3).

We may also apply process tracing in the empirical analysis of critical junctures and path dependence. In combination with diachronic comparative case design which is feasible to establish and explain continuity and change in caveats patterns over time (e.g., Husby 2015), the close-up and high-resolution method of process tracing can make an important contribution in providing more precise measurement of critical junctures and tipping points as regards caveats-policies in individual cases (Tarrow 1995: 474). While the study of path dependence details causal mechanisms that reinforce early predominance (Hall 2007: 92), Levy argues that the high resolving power of process-tracing research is also capable of revealing the precise timing of critical junctures and turning-points as well as the critical mechanisms causally involved in process (2008: 12). Whether used for theory-developing or theory-testing purposes, process tracing is likely to improve our understanding of both trigger-mechanisms and reinforcing mechanisms at work in foreign policy-making processes related to caveats behavior.

It is significant in what sequence we combine QCA and process-tracing methodologies. For theory-building purposes, we suggest single-case process tracing ("practice tracing") be given precedence to any multiple case-study design. However, for theory testing, we are better served by first executing QCA to be able to uncover the extent of "multiple conjectural causation" (Ragin 1987). If we were to find that three different clusters or combinations of INUS conditions are involved in producing the same outcome (caveats or no caveats) or different kinds of caveats, we would subsequently conduct one process-tracing study at the level of the single case for each one of the three groups of cases identified in the preceding QCA to trace the three distinct causal paths/mechanisms in question. Conducting a single-case process-tracing procedure before any QCA would risk us overlook alternative causal paths to the same outcome. The best practice would be to use QCA to get a crude overview of the causal landscape, and then apply process tracing at the level of the single case to get a high-resolution picture of the respective causal paths from input factors to the outcome. For our purposes, the tracing procedure would include the detailed study of the several phases of the foreign policy-making process explained in Chapter 6.

References

Adler, E. (1997). Seizing the Middle Ground: Constructivism in World Politics. *European Journal of International Relations, 3*(3), 319–363.

Allison, G., & Zelikow, P. (1999). *Essence of Decision. Explaining the Cuban Missile Crisis.* New York, NY: Longman.

Almond, G., & Verba, S. (1963). *The Civic Culture: Political Attitudes and Democracy in Five Nations.* Thousand Oaks, CA: Sage.

Aristotle. (1979 [c.350 BCE]). *Politics.* Harmondsworth: Penguin.

Auerswald, D. P., & Saideman, S. M. (2014). *NATO in Afghanistan: Fighting Together, Fighting Alone.* Princeton, NJ: Princeton University Press.

Bartlett, L., & Vavrus, F. (2017). *Rethinking Case Study Research. A Comparative Approach.* Oxford, UK: Routledge.

Bates, R., Weingast, B., Greif, A., Levi, M., & Rosenthal, J.-L. (1998). *Analytic Narratives.* Princeton, NJ: Princeton University Press.

Bauman, Z. (1978). *Hermeneutics and Social Science: Approaches to Understanding.* London: Routledge.

Bennett, A., & Checkel, J. T. (Eds.). (2015). *Process Tracing. From Metaphor to Analytical Tool.* Cambridge: Cambridge University Press.

Bennett, A., Lepgold, J., & Unger, D. (1994). Burden-Sharing in the Persian Gulf War. *International Organization, 48*(1), 39–75.

Blaikie, N. (2009). *Designing Social Research*. Hoboken, NJ: Wiley.

Brady, H. E., & Collier, D. (Eds.). (2010). *Rethinking Social Inquiry. Diverse Tools, Shared Standards*. New York, NY: Rowman & Littlefield.

Brady, H. E., Collier, D., & Seawright, J. (2010). Refocusing the Discussion of Methodology. In H. E. Brady & D. Collier (Eds.), *Rethinking Social Inquiry. Diverse Tools, Shared Standards* (pp. 15–32). New York, NY: Rowman & Littlefield.

Brams, S. J. (1994). *Theory of Moves*. Cambridge: Cambridge University Press.

Brinton, C. (1965). *The Anatomy of Revolution*. New York, NY: Vintage Books.

Bryman, A. (2001). *Social Research Methods*. Oxford: Oxford University Press.

Byrne, D., & Ragin, C. C. (Eds.). (2013). *The Sage Handbook of Case-Based Methods*. London: Sage.

Checkel, J. T. (2006). Tracing Causal Mechanisms. *International Studies Review, 8*(2), 362–370.

Collier, D. (1993). The Comparative Method. In A. W. Finifter (Ed.), *Political Science: The State of the Discipline II* (pp. 105–120). Washington, DC: American Political Science Association.

Collier, D. (1999, Winter). Data, Field Work and Extracting New Ideas at Close Range. *APSA-CP Newsletter*, pp. 1–6.

Cronqvist, L., & Berg-Schlosser, D. (2009). Multi-value QCA (mvQCA). In B. Rihoux & C. C. Ragin (Eds.), *Configurational Comparative Methods. Qualitative Comparative Analysis and Related Techniques* (pp. 69–86). Los Angeles, CA: Sage.

de Mesquita, B. B. (2000). *Principles of International Politics: People's Power, Preferences, and Perceptions*. Washington, DC: QC Press.

Eckstein, H. (1975). Case Study and Theory in Political Science. In F. I. Greenstein & N. W. Polsby (Eds.), *Handbook of Political Science* (Vol. 7, pp. 119–161). Reading, MA: Addison-Wesley.

Eisenhardt, K. M. (1989). Building Theories from Case Study Research. *The Academy of Management Review, 14*(4), 532–550.

Elman, C. (2005). Explanatory Typologies in Qualitative Studies of International Politics. *International Organization, 59*(2), 293–326.

Fearon, J. D. (1998). Domestic Politics, Foreign Policy, and Theories of International Relations. *Annual Review of Political Science, 1*(3), 289–313.

Fermann, G. (1988). *UNEF II – 1973–89: Instrument for forhandlet konflik-tløsning* (NUPI Report No. 120). Oslo: Norwegian Institute for International Affairs.

Fermann, G. (1992). *Internasjonal fredsbevaring 1956–1990. En sammenliknende undersøkelse. Forsvarsstudier 5/1992*. Oslo: Institutt for forsvarsstudier. https://brage.bibsys.no/xmlui/bitstream/handle/11250/99313/4/FS0592.pdf.

Fermann, G. (Ed.). (2013). *Utenrikspolitikk og norsk krisehåndtering.* Oslo: Cappelen Damm Akademika. https://www.cappelendamm.no/_utenrikspolitikk-og-norsk-kriseh%C3%A5ndtering-gunnar-fermann-9788202378691.

Frendreis, J. P. (1983). Explanation of Variation and Detection of Covariation. The Purpose of Logic and Logic of Comparative Analysis. *Comparative Political Studies, 16*(2), 255–272.

Frost-Nielsen, P. M. (2016). *Betingede forpliktelser. Nasjonale reservasjoner i militære koalisjonsoperasjoner.* Ph.D. dissertation in Political Science, Department of Sociology and Political Science, Norwegian University of Science and Technology (NTNU), Trondheim.

Frost-Nielsen, P. M. (2017). Conditional Commitments: Why States Use Caveats to Reserve Their Efforts in Military Coalition Operations. *Contemporary Security Policy, 38*(3), 371–397.

George, A. L. (1979). Case Studies and Theory Development: The Method of Structured Focused Comparison. In P. G. Lauren (Ed.), *Diplomatic History: New Approaches* (pp. 43–68). New York, NY: Free Press.

George, A. L., & Bennett, A. (2005). *Case Studies and Theory Development in Social Sciences.* Cambridge, MA: MIT Press.

Gerring, J. (2007). *Case Study Research. Principles and Practices.* New York, NY: Cambridge University Press.

Hague, R., Harrop, M., & Breslin, S. (Eds.). (1998). *Government and Politics: An Introduction.* Basingstoke: Macmillan.

Hall, J. R. (2007). Historicity and Socio-historical Research. In W. Outhwaite & S. P. Turner (Eds.), *The Sage Handbook of Social Science Methodology* (pp. 82–101). London: Sage.

Haney, P. J. (2002). *Organizing for Foreign Policy Crises: Presidents, Advisers, and the Management of Decision-Making.* Ann Arbor, MI: University of Michigan Press.

Harris, W. A. (1997). On "Scope Conditions" in Sociological Theories. *Social and Economic Studies, 46*(4), 123–127.

Henriksen, Dag. (2007). *NATO's Gamble. Combining Diplomacy and Airpower in the Kosovo Crisis 1998–1999.* Annapolis, MD: Naval Institute Press.

Hudson, V. M. (2005). Foreign Policy Analysis: Actor-Specific Theory and the Ground of International Relations. *Foreign Policy Analysis, 1*(1), 1–30.

Hudson, V. M. (2007). *Foreign Policy Analysis: Classical and Contemporary Theory.* Boulder, CO: Rowman and Littlefield.

Husby, G. (2015). *Fra hull i luften, til hull i Gaddafis bunker. Bruk av politiske reservasjoner på norsk militærmakt i flernasjonale koalisjonsoperasjoner. En komparativ studie av F-16 bidragene i Kosovo, Afghanistan og Libya.* Master Thesis in Political Science, Department of Sociology and Political Science. Norwegian University of Science and Technology (NTNU), Trondheim.

Ingesson, T. (2016). *The Politics of Combat: The Political and Strategic Impact of Tactical-Level Subcultures, 1939–1995.* Lund: Ph.D. Dissertation, Faculty of Social Sciences and Department of Political Science, Lund University. http://portal.research.lu.se/ws/files/7253766/Tony_Ingesson_Politics_of_Combat.pdf.

Ingesson, T. (2017). Trigger-Happy, Autonomous, and Disobedient: Nordbat 2 and Mission Command in Bosnia. *The Strategy Bridge,* 20 September. https://thestrategybridge.org/the-bridge/2017/9/20/trigger-happy-autonomous-and-disobedient-nordbat-2-and-mission-command-in-bosnia.

Johnson, R. B., Onwuegbuzie, A. J., & Turner, L. A. (2007). Toward a Definition of Mixed Methods Research. *Journal of Mixed Methods Research, 1*(2), 112–133.

Kawulich, B. B. (2005). Participant Observation as a Data Collection Method. *Qualitative Social Research, 6*(2). http://www.qualitative-research.net/index.php/fqs/article/view/466/9*96.

Kehr, E. (2012 [1927]). *Der Primat der Innenpolitik. Gesammelte Aufsätze zur preußisch-deutschen Sozialgeschichte im 19. und 20. Jahrhundert.* Berlin: Walter De Gruyter & Co.

King, G., Keohane, R. O., & Verba, S. (1994). *Designing Social Inquiry. Scientific Inference in Quantitative Research.* Princeton, NJ: Princeton University Press.

Kögler, H.-H. (2011). Understanding and Interpretation. In W. Outhwaite & S. P. Turner (Eds.), *The Sage Handbook of Social Science Methodology* (pp. 363–383). London: Sage.

Levy, J. S. (2002). Qualitative Methods in International Relations. In M. Brecher & F. P. Harvey (Eds.), *Millennial Reflections on International Studies* (pp. 432–454). Ann Arbor, MI: University of Michigan Press.

Levy, J. S. (2007). Theory, Evidence, and Politics in the Evolution of Research Programs. In R. N. Lebow & M. I. Lichbach (Eds.), *Theory and Evidence in Comparative Politics and International Relations.* New York, NY: Palgrave Macmillan.

Levy, J. S. (2008). Case Studies: Types, Designs, and Logics of Inference. *Conflict Management and Peace Science, 25*(1), 1–18.

Lieberson, S. (1992). Small N's and Big Conclusions: An Examination of the Reasoning in Comparative Studies Based on a Small Number of Cases. In C. C. Ragin & H. S. Becker (Eds.), *What Is a Case? Exploring the Foundations of Social Inquiry* (pp. 105–118). Cambridge: Cambridge University Press.

Lijphart, A. (1971). Comparative Politics and Comparative Method. *American Political Science Review, 65*(3), 682–698.

Lijphart, A. (1975). The Comparable-Cases Strategy in Comparative Research. *Comparative Political Studies, 8*(2), 158–177.

Lombardi, B. (2008). All Politics Is Local: Germany, the Bundeswehr, and Afghanistan. *International Journal, 63*(3), 587–605.

Machiavelli, N. (1961 [1531]). *The Prince*. Harmondsworth: Penguin.

Mackie, J. L. (1965). Causes and Conditions. *American Philosophical Quarterly, 4*(2), 245–264.

Mill, J. S. (2002 [1891]). *A System of Logic*. Honolulu; HI: University Press of the Pacific.

Mitchell, J. Clyde. (1983). Case and Situational Analysis. *Sociological Review, 31*(2), 187–211.

Mjøset, L. (2009). The Contextualist Approach to Social Science Methodology. In D. Byrne & C. C. Ragin (Eds.), *Case-Based Methods* (pp. 39–68). London: Sage.

Møller, J. (2015). *Statsdannelse, regimeforandring og økonomisk udvikling*. Copenhagen: Hans Reitzels Forlag.

Moore, B. (1966). *Social Origins of Dictatorship and Democracy*. Boston, MA: Beacon Press.

Moses, J. W., & Knutsen, T. L. (2012). *Ways of Knowing. Competing Methodologies in Social and Political Science*. London: Palgrave Macmillan.

Olson, M. (1971). *The Logic of Collective Action: Public Goods and the Theory of Groups*. Cambridge, MA: Harvard University Press.

Olson, M., & Zeckhauser, R. (1966). An Economic Theory of Alliances. *The Review of Economics and Statistics, 48*(3), 266–279.

Outhwaite, W., & Turner, S. P. (Eds.). (2011). *The Sage Handbook of Social Science Methodology*. London: Sage.

Platt, J. (2007). Case Study. In W. Outhwaite & S. P. Turner (Eds.), *The Sage Handbook of Social Science Methodology* (pp. 102–127). London: Sage.

Pouliot, V. (2015). Practice Tracing. In A. Bennett & J. T. Checkel (Eds.), *Process Tracing. From Metaphor to Analytical Tool* (pp. 102–127). Cambridge: Cambridge University Press.

Przeworski, A., & Teune, H. (1970). *The Logic of Comparative Social Inquiry*. New York, NY: Wiley.

Ragin, C. C. (undated). *What Is Qualitative Comparative Analysis?* Tucson, AZ: University of Arizona. http://eprints.ncrm.ac.uk/250/1/What_is_QCA.pdf.

Ragin, C. C. (1987). *The Comparative Method: Moving Beyond Qualitative and Quantitative Strategies*. Oakland, CA: University of California Press.

Ragin, C. C. (1994). *Constructing Social Research. The Unity and Diversity of Method*. Thousand Oaks, CA: Pine Forge Press.

Ragin, C. C. (2000). *Fuzzy-Set Social Science*. Chicago, IL: University of Chicago Press.

Ragin, C. C. (2007). Comparative Methods. In W. Outhwaite & S. P. Turner (Eds.), *The Sage Handbook of Social Science Methodology* (pp. 67–81). London: Sage.

Ragin, C. C. (2008). *Redesigning Social Inquiry: Fuzzy Sets and Beyond*. Chicago, IL: University of Chicago Press.

Ragin, C. C. (2009). Qualitative Comparative Analysis Using Fuzzy Sets (fsQCA). In B. Rihoux & C. C. Ragin (Eds.), *Configurational Comparative Methods. Qualitative Comparative Analysis and Related Techniques* (pp. 87–122). Los Angeles, CA: Sage.

Ragin, C. C., & Becker, H. S. (Eds.). (1992). *What Is a Case? Exploring the Foundations of Social Inquiry*. Cambridge: Cambridge University Press.

Rihoux, B. (2013). QCA 25 Years After "The Comparative Method": Mapping, Challenges, and Innovations—Mini-Symposium. *Political Research Quarterly, 66*(2), 167–235.

Rihoux, B., & De Meur, G. (2009). Crisp-Set Qualitative Comparative Analysis. In B. Rihoux & C. C. Ragin (Eds.), *Configurational Comparative Methods. Qualitative Comparative Analysis and Related Techniques* (pp. 33–68). Los Angeles, CA: Sage.

Rihoux, B., & Ragin, C. C. (Eds.). (2009). *Configurational Comparative Methods. Qualitative Comparative Analysis and Related Techniques*. Los Angeles, CA: Sage.

Rose, G. (1998). Neoclassical Realism and Theories of Foreign Policy. *World Politics, 51*(1), 144–172.

Rosenau, J. N. (Ed.). (1974). *Comparing Foreign Policies: Theories, Findings, and Methods*. New York, CA: Wiley.

Rynning, S., & Guzzini, S. (2002). Realism and Foreign Policy Analysis. In F. Charillon (Ed.), *Politique etrangere: Nouveaux regards*. Paris: Universitaires de France.

Sambanis, N. (2004). What Is Civil War? Conceptual and Empirical Complexities of an Operational Definition. *Journal of Conflict Resolution, 48*(6), 814–858.

Schneider, S. L. (2007). Experimental and Quasi-experimental Designs in Behavioral Research: On Context, Crud, and Convergence. In W. Outhwaite & S. P. Turner (Eds.), *The Sage Handbook of Social Science Methodology* (pp. 172–189). London: Sage.

Simons, H. (2009). *Case Study Research in Practice*. London: Sage.

Skocpol, T. (1979). *States and Social Revolutions: A Comparative Analysis of France, Russia, and China*. New York, NY: Cambridge University Press.

Smelser, N. J. (1973). The Methodology in the Social Sciences. In D. P. Warwick & S. D. Osherson (Eds.), *Comparative Research Methods* (pp. 42–86). Englewood Cliffs, NJ: Prentice-Hall.

Smith, S., Hadfield, A., & Dunne, T. (Eds.). (2008). *Foreign Policy: Theories, Actors, Cases*. Oxford: Oxford University Press.

Snyder, G. H. (1984). The Security Dilemma in Alliance Politics. *World Politics, 36*(4), 461–495.

Stake, R. E. (1995). *The Art of Case Study Research*. London: Sage.

Tarrow, S. (1995). The Europeanisation of Conflict: Reflections from a Social Movement Perspective. *West European Politics, 18*(2), 223–251.

Thies, C. G., & Breuning, M. (2012). Integrating Foreign Policy Analysis and International Relations Through Role Theory. *Foreign Policy Analysis, 8*(1), 1–4.

Thomas, G. (2011). A Typology of the Case Study in Social Science Following a Review of Definition, Discourse, and Structure. *Qualitative Inquiry, 17*(6), 511–521.

Van Evera, S. (1997). *Guide to Methods for Students of Political Science.* Itacha, NY: Cornell University Press.

Vennesson, P. (2008). Case Study and Process Tracing: Theories and Practices. In D. Della Porta & M. Keating (Eds.), *Approaches and Methodologies in the Social Sciences. A Pluralist Perspective* (pp. 223–239). Cambridge: Cambridge University Press.

Wicks, D. (2012). Deviant Case Analysis. In A. J. Mills, G. Durepos, & E. Wiebe (Eds.), *Encyclopedia of Case Study Research* (pp. 290–291). London: Sage.

Wieviorka, M. (1992). Case Studies: History or Sociology? In C. C. Ragin & H. S. Becker (Eds.), *What Is a Case? Exploring the Foundations of Social Inquiry* (pp. 159–172). Cambridge: Cambridge University Press.

Wolf, E. R. (1968). *Peasant Wars of the Twentieth Century.* New York, NY: Harper & Row.

PART VI

Conclusions

CHAPTER 12

Recapitulations and Contributions

The use of caveats on national contributions to coalition forces is likely as old as organized warfare and alliance politics. The issue of national reservations on the use of force rose to political prominence in defense and policy circles with NATO's campaign in Afghanistan. The application of caveats in the context of UN, NATO, and "Coalition-of-the willing" operations after the Cold War have stimulated research on the politics of caveats. Why do states make substantial military contributions to coalition operations, while at the same time apply reservations, caveats, to how the coalition can use the national military contributions? Caveats often signal reluctant participation and is a challenge to the efficient use of coalition forces and implementation of political mandates.

On this backdrop, we have offered an empirical research program for the study of the politics of caveats. In Social Science, the essence of an empirical research program is to provide directions on *how* to study some social or political phenomenon. This function implies that the outlined program *by itself* does not pretend to provide any conclusive answers to substantial research questions such as; how prevalent is the phenomenon of caveats in post-Cold War coalition operations and deeper down in history? What are the conditions for the application of caveats? What mechanisms are involved in the explanation of patterns of caveats in foreign policy and coalition operations? What are the effects and impacts of caveats-use on coalition forces and alliance politics? Would more states abstain from contributing to coalitions if caveats were not an option?

© The Author(s) 2019   237
G. Fermann, *Coping with Caveats in Coalition Warfare*,
https://doi.org/10.1007/978-3-319-92519-6_12

It is mainly for subsequent research to provide more conclusive answers to such empirical research questions.

We have argued that the contribution of the empirical research program rather be assessed on its capacity to inspire, direct, and facilitate research. We suggest the program assessed on its capacity to reason precise conceptual constructs capable of delimiting the phenomenon of caveats consistently. This requirement includes the conceptual capacity to distinguish between different expressions of national reservations on the use of force in coalition warfare when observed. Furthermore, we propose the program judged on the coherence and the capacity of its analytical framework to guide the selection of theory by which arguments are reasoned and empirical propositions deduced. Finally, the utility of the research program depends on the usefulness of the directions for the gathering and analyses of data, while taking into account the particular attributes of the research field in question, as well as the ontological content of the analytical framework suggested. The program needs to deliver usefully on all these accounts. To this, we may add that the empirical research program also be assessed on its capacity to inspire new and interesting research questions on the politics of caveats.

The programmatic intent of the study has been encouraged by a triple recognition. First, caveats as national reservations on the use of force in coalition operations is a real phenomenon that tends to erode the military efficiency of coalition forces. This is the reason why national reservations on the use of force in coalition operations have become a contentious issue in alliance politics. After the Cold War, the use of caveats in coalition operations has been discussed at the highest political level in The United Nations and NATO (United Nations 2000; NATO 2006). Force Commander's have criticized the application of caveats as a threat to the integrity of coalition forces (Jones 2004; Johnson 2004; Clark 2001). For such reasons, national reservations on the use of force in coalition operations warrant scholarly attention. There are some real epistemological puzzles to solve and a host of potential research questions of political relevance to pursue. Not least related to the *foreign* politics of caveats.

Second, there is an emerging academic literature and a scholarly dialogue to build and reflect upon. In particular, I have benefitted from the several years of collaboration with my former master and PhD student Per Marius Frost-Nielsen (2009, 2011, 2013, 2016, 2017, 2018, forthcoming) in researching of the present study. Responding to the significant operational and political challenges national reservations on the use of force pose to military coalitions, caveats have frequently been mentioned in the scholarly security literature since the early 2000s (Auerswald 2004). In particular, references to caveats are found in research on

1. civil-military relations in complex military operations (Auerswald and Saideman 2014; Ruffa et al. 2013);
2. counter-insurgency doctrine and state-building activities (Meyer 2013);
3. combat-effectiveness and multinational operations (Marten 2007; Deni 2004);
4. multilateral military intervention and burden-sharing (Richter and Webb 2014; Kreps 2008); and
5. the comparative study of democratic participation in armed conflict (Mello 2014).

The designs of some of the best research in the field are to a considerable extent shaped by approaches from Comparative Politics (CP) and International Relations (IR). We have also argued that these approaches tend to have less to say about actual decision-making processes and the theoretical implications of seeing national reservations on the use of force as a foreign policy instrument (Fermann 2013). Moreover, scholarship on the politics of caveats fails to agree on how caveats are defined and operationalized for empirical research (Frost-Nielsen 2016: 5, 44). This particular state of affairs is an obstacle to systematic research, tends to slow down the longer-term development of reliable databases on which high N-research depends, and thus counteract generic and cumulative research ambitions.

Third, caveats-research is likely to benefit from the systematic application of the approach of Foreign Policy Analysis (FPA). FPA invites the scholar to consider the nature of foreign policy-making; understand the implications of seeing national reservations on the use of force as a foreign policy instrument of the state; grasp the causal mechanisms of foreign policy-making and implementation; and capture the interaction between explanatory factors at different levels of analyses that largely remains to research. In these regards, the FPA approach to caveats has the potential to supplement existing research by shedding light on tracks seldom or lightly traveled by scholars using CP and IR-approaches as points of departure for political caveats-research.

In the final chapter of the study, we recapitulate the main conceptual, analytical, theoretical, deductive, and methodological contributions to the empirical research program on national reservations on the use of force. We acknowledge that considerable deductive effort remains to clarify the theoretical mechanisms at work in several of the hypotheses suggested. Furthermore, several empirical propositions are formulated as macro hypotheses, thus leaving it to subsequent researchers to make decisions on exactly how to operationalize the independent variables in question. Finally, a lot of nitty-gritty work remains as to exactly how single and comparable-cases designs are to be utilized for theory-building and theory-testing purposes. Then again, an empirical research program is no instruction manual. The study is limited to providing clues and directions, which competent researchers would need and want to specify for their own particular research purposes.

CONCEPTUAL CONTRIBUTIONS: DELIMITING THE EMPIRICAL FOOTPRINT OF THE CONCEPT OF CAVEATS

The *first main contribution* of the research program relates to the conceptual identification of the phenomenon of caveats. Before constructing a framework for analysis that is capable of better explaining *why* and *how* national reservations on the use of force is applied in coalition contexts, it is crucial to have a clear grasp of *what* precisely caveats are. We require a concept that makes the phenomenon of caveats recognizable, possible to measure, and capable of producing reliable data on caveats as dependent or independent variable. The conception of national reservations on the use of force also needs to distinguish clearly between different kinds of caveats.

In this effort, the literature on the politics of caveats is both suggestive and confusing. How can we systematically compare reports on and studies of coalitions and contributing states on the prevalence of caveats when we have yet to agree upon the conceptual and operational delimitation of the phenomenon? Because studies refer to only partially, overlapping phenomena, are almost unspoken about some important conceptual properties, and are either too broad or too narrow in their conception of caveats to capture the essence or complexity of the phenomenon, any attempt to put together a systematic database on caveats is premature or suggestive.

In reviewing the literature on the politics of caveats, we found some studies to refer to caveats in terms of how national reservations on the use of force have prevented military units to participate in offensive and risky military operations (Mello 2014: 113–114; Ringsmose 2010: 328; Sky 2007: 16). Other contributors were focusing on the gate-keeping functions of national red flag-holders and staff officers who in different ways intervene in the coalition's chain of command to satisfy the political concerns of some national government (Saideman and Auerswald 2012: 69–70; Høiback 2009: 23–34; Young 2003: 115).

In some studies, researchers conceptually link caveats to national constitutional conditions that lead to the reserved coalition-behavior (e.g., van der Meulen and Kawano 2008; Koschut 2014: 351–354). One study allows caveats to cover the whole range of financial, logistical-, and capacity-related restrictions regarding the military robustness of the contingent (Brophy and Fisera 2010: 1). A final conception of national reservations on the use of force, especially in the context of the NATO campaign in Afghanistan, is the geographical limitations on force mobility states impose on their military contingents (Kay 2013: 109–110; Noetzel and Rid 2009: 75; Noetzel and Schreer 2009: 532).

The inconsistent use of the caveats concept across scholarships implies that it is problematic to compare and lump together the ISAF coalition staff-finding that some 70 instances of caveats were at work in Afghanistan (Bergen 2011: 189) with, say, Otto Trønnes' (2012) finding that the Norwegian government applied tenfold with caveats to its contributions to the multinational war effort in Afghanistan from 2001 through 2008. Obviously, The Norwegian government did not by itself account for more than half of the total of coalition caveats in Afghanistan. Also, the conception of caveats Per Marius Frost-Nielsen (2016: 10–19) applies to research on Danish, Dutch, and Norwegian caveats-use in the 2011 Libya coalition operation (2017: 3–4), deviates considerably from the notion of caveats Ben Lombardi (2008) uses in the empirical mapping of German caveats-policies in Afghanistan.

How can we rectify this state of affairs? The lessons we learned from the conceptually fragmented literature was, one, that any empirical research program on the politics of caveats need to establish a well-reasoned and measurable conception of caveats at the outset that takes into consideration the best of what the caveats-literature has to offer. In some instances, scholars cross the border to an adjacent phenomenon. More often, the lack of compatibility are due to scholars emphasize different aspects of the phenomenon of caveats. Second, we need to liberate the concept of caveats from its contextual particularities in time and space to transcend the local experience (Platt 2007: 120). Finally, the conceptualization should be extensive enough to grasp the complexity of the phenomenon. However, the scope of the concept should also be narrow enough to exclude related, but different phenomena such as the decision to contribute militarily to coalition forces, other instruments of foreign policy-making (Gerring 1999; Fermann 2013: 72–83), other constitutional checks on the use of force abroad (Mello 2014: 6), and other ways of imposing national control on the use of force in coalition operations (Auerswald and Saideman 2014: 5–12).

The first step to identify the phenomenon in question was to define caveats in terms of *national reservations on the use of force in coalition contexts* (Frost-Nielsen 2016: 12–13). By so doing, we distinguished caveats from other adjacent phenomena such as the decision whether to participate in the coalition and the decision to offer a substantial military unit rather than merely a symbolic presence. Caveats thus relate to nation-specific conditions for the use of force in the field—when, how, to what extent, and where within the area of deployment of the coalition.

To further distinguishing caveats from other adjacent phenomena that are likely to have different causes and consequences, the second step was to specify a set of properties or criteria such national reservations on the use of force need to fulfill to qualify as caveats. Crucially, we reserved the concept of caveats for *self-imposed national reservations on the use of force that are the result of politically motivated and calculated decision-making*. The importance of emphasizing the political nature of caveats is that it captures the reality of caveats as a foreign policy instrument that reflect some national priorities. This political understanding of caveats distinguishes the concept from restrictive military behavior that results from uncoordinated action and incompetence due to chance, or some technical, logistical, managerial, or financial limitation (Findlay 2002: 354–359). In reviewing the literature, we found that some force-behavior that is due to lack of coordination and other kinds of resource-limitations are easily mistaken for caveats (Frost-Nielsen 2016: 15).

Furthermore, the concept of caveats as national reservations on the use of force is limited to *military contingents subordinated to a unified chain of command*, and to military conduct *relating to some common mechanism for the regulation of the use of force*. This dual specification draws a line against supportive operational contributions such as the facilitation of military hospitals and supply lines. More obvious, the specification also rules out unilateral military operations from the empirical universe of caveats. However, the analytical condition that national contingent subordinated to a coalition chain of command also implies an expansive element in that it is not limited to the inclusion of UN, NATO, and "Coalition-of–the-Willing" operations after the end of the Cold War. Thus, we extended the generic scope and value of the concept to include all past, present, and future coalition forces that meet said criteria. Finally, the specification anticipated two essential ingredients in the subsequent arguing of caveats empirical indicators related to the extent coalition command overruled by national representatives in the chain of command, and to national behavioral deviations from the use of force permitted for in coalition rules of engagement (RoE).

A further specification of the definition of caveats was the decision to *include not only national reservations that are documented formally in political statements and admitted in particular operational codes of conduct, but also informal, undeclared, and even unadmitted use of caveats that show in actual behavior*. The study of behavioral practice-patterns will reveal the concealed use of caveats practiced by some countries in ISAF, and is particularly important when the nature of military operations changes (Auerswald and Saideman 2014: 146–147). The change will potentially affect states' political views on the operation, and, in turn, how they assess the political feasibility of caveats.

Finally, we drew on Per Marius Frost-Nielsen's (2016: 15–16) line of reasoning to argue a *symmetrical understanding of the term "reservations" to allow for both restrictive and permissive interpretations of the caveats-phenomenon*. The literature indicates that the large majority of national reservations on the use of force are restrictive. Still, the literature review showed a handful of reservation instances that were of a permissive kind. For instance, the Dutch contingent in ISAF reserved itself the right to use their air support unit even when vetoed by coalition command (Auerswald and Saideman 2014: 166). Furthermore, we may as well classify the regular practice of the United States to insist on the prerogative of having an American general lead the coalition force as a instance of permissive caveat. This final specification of the theoretical definition of caveats represents an extension of the concept.

By further specifying the initial definition of caveats as national reservations on the use of force, the definition evolved into a conceptual construct with considerable higher resolving and phenomena-discriminating power. By now, caveats was much more precisely defined as *politically motivated, national conditions for the use of force in a coalition force where military contingents are subordinated to a unified chain of command and relate to some common regulation of the use of force. Particular national conditions for the use of force can be either of a restrictive, or a permissive kind, and may be formally recognized as such, or be informal, undeclared, and even unadmitted by the force-contributing state, only to be observed in actual force-deviating behavior not related to lack of capacity or coordination.*

The final step in the effort to pin down the phenomenon of caveats and prepare the concept for empirical research was to specify how to actually measure the concept. Following several lines of reasoning, we concluded that the empirical footprint of caveats (operational definition) is observable along three dimensions. The concept of caveats be measured;

1. as national deviations from the force-regulating coalition rules of engagement (RoE) in terms of when use of force is permitted and how on any level of operational command
2. in national interference in the coalition chain of command in discretion to veto orders by means of designated red card-holder, staff officers assigned to coalition command, or personnel further down the coalition chain of command, and
3. in national conditions on the extent to which coalition is delegated the authority to make full use of the operational capacity of the national contingent as to where, when, and how contingent be deployed and used in theater of war.

Arguably, much of the conceptual confusion in the literature originates from the fact that in many studies only one of the several operational dimensions are applied, and in some cases inconsistently so. By systematically applying the more complex operational definition, tapping into the mechanisms regulating the use of force in coalition operations, we may observe and reflect upon previously undetected instances and kinds of caveats. Common to all three operational dimensions is the fundamental attribute that national reservations on the use of force are not the reflection of some lack of military capacity, insufficient coordination, or chance, but the result of a calculated political decision, serving some foreign policy-purpose. However, in classifying particular caveats, it is, depending on the research question, crucial to consult also other distinguishing properties of the concept relating to whether the caveats in question are of a restrictive or permissive kind, and the extent to which caveats used are officially recognized and admitted.

In a string of single and comparative case studies, Per Marius Frost-Nielsen (2016, 2017) has utilized a similar compound conceptualization of caveats in studies of Norwegian, Danish and Dutch coalition participation in Afghanistan and Libya. Frost-Nielsen's case studies also serve as a plausibility probe (Eckstein 1975: 108–113) in that he also uses the empirical opportunity of the case study design to assess the validity and usefulness of the caveats concept as defined and operationalized above. Crucial indications of a fruitful conceptual construct (a variable) are that the concept is mirror imaged in actual empirical observations and that the phenomenon in question varies across units, space, and time. Frost-Nielsen's research does indeed confirm the usefulness of the concept of caveats as national conditions for the use of coalition forces in these regards. In particular, the variation on the dependent variable is observable across cases, states, and coalition forces. In a partly overlapping operationalization of caveats, Auerswald and Saideman's study of ISAF in Afghanistan (2014: 19–20) documented variation on the dependent variable as well. Different states applied caveats to a varying extent.

A final indication of the usefulness of the conceptual contribution of the study is the capacity to distinguish between different classes and types of caveats. As defined, the conception of caveats as national reservations on the use of force in coalition forces comes in many shapes. Restrictive caveats show in the unwillingness of national contingents to fight when the common RoE allows or directs them to apply use of force. It shows in the practice of having a national representative—a red card-holder—acting as a gate-keeper or veto-player in the coalition chain of command for missions and tasks the government does not want to take on. Restrictive caveats also show in the unwillingness to fight and contribute outside the assigned area of deployment, or at night, even if the operational threat-situation so requires and the Force Commander so orders. Restrictive caveats furthermore show in a defensive posture when offensive action is required and allowed for in coalition RoE. Permissive caveats, on the other hand, appears in the alliance leader's insistence to take on the military leadership of the coalition, and in a participating state's willingness to use excessive force when the contingent is in dire straits.

The multidimensional operationalization of national reservations on the use of force argued in the present study thus invites research explaining why coalition members may choose to apply particular kinds of caveats, officially admitted or not. The main challenge for empirical research is, of course, to document that informal or officially unrecognized caveats are at work and to render probable that the reservations in question are indeed politically motivated.

ANALYTICAL CONTRIBUTIONS: REASONING THE FRAMEWORK FOR ANALYSIS—APPROACH, THEORIES, AND HYPOTHESES

The *second main contribution* of the empirical research program was the elaboration of a *framework for analysis* for the study of the politics of national reservations on the use of force. Seen as a whole, the purpose of any framework for analysis is to prepare for and facilitate the testing of theory by empirically analyzing hypotheses. We elaborated the particular framework for the study of the politics of caveats through three distinct steps in a dual inductive-deductive move that started with;

1. the reasoning of an *approach* providing direction and coherence to research;
2. continued with the selection and review of several bodies of *theory* on three-four levels of analyses from which explanations of caveats-behavior be reasoned; and finally,
3. resulted in the actual deduction and formulation of *hypotheses* on the conditions for national coalition-behavior, in particular concerning national reservations on the use of force.

The *first step* in the development of the framework for analysis was to argue the approach of *Foreign Policy Analysis* (FPA) as a promising point of departure for the systematic study of caveats-behavior. In FPA, we frame caveats as a foreign policy instrument facilitating the projection of national interests and issue-specific priorities. Following FPA, the primary agency in the political study of caveats rests with the foreign policy-making state. However, we do not restrict ourselves to the perception of the state as a unitary actor acting on some simple notion of rationality. Rather, we see the state as a social organization made up of multiple decision-making and implementing institutions and actors that may both be sensitive toward external environments and proactive in behavior. We argued that the foreign policy decision-making process is institutionalized and embedded in collective self-conceptions, culture, and norms influencing policy decisions on foreign policy preferences, means, and behavior.

Although the state is considered politically independent (sovereign), its freedom of action in international affairs is limited by other states, multilateral institutions, and transnational actors operating within the political, economic, and normative structures of global politics. We also insisted that foreign policy-making on issues related to alliance politics and reserved use of force in coalition operations is likely to be influenced by society and domestic politics. In FPA, this implies that key decision-makers' perceptions and narratives of the dual environments strongly influence what is deemed politically feasible to do in foreign policy, including decisions related to alliance dynamics and the participation in coalition forces. In this sense, foreign policy-making on the issue of caveats is a rather complex decision-process pinched between a rock (society, domestic politics) and a hard place (global politics).

Still, there is a space for political engineering in foreign policy-making, according to FPA. Key decision-makers engage in political, creative action in a calculated manner, with the aim of exploiting the scope for action offered by structural opportunities and the actions or passivity of other actors. In this context, we argued that caveats is a political instrument that foreign policy-makers may use to extend the scope for political maneuvering (SPM), and make the military participation in coalition forces politically feasible domestically. Preparing for the application of theories on the implementation of policies, we also argued that caveats be used to impose stricter national control of military contingents in coalition warfare on the very reasonable assumption that "national political control of military does not stop when coalition wars start" (Saideman 2018).

The FPA approach has been criticized for only qualifying as a "pretheory," implying that FPA *by itself and on its own* lacks the required ontological content from which empirical propositions on the explanation of foreign policy phenomena can be deduced (Rosenau 1966). This is true, but only up to a point. The ontological deficit of FPA does not render this particular approach and other metaphors, taxonomies, and major approaches useless to give direction to the choice of middle-range theories from which to deduce hypotheses. We used the distinction Imre Lakatos (1978: 47–51) made between the "hard core" (the approach) and the "auxiliary hypotheses" (theories) of a scientific research program (Caldwell 1991) to clarify the epistemological division of labor between the approach of FPA, and the middle-range theories from which hypotheses can be inferred about the application of national reservations on the use of force: Typically, approaches ("pre-theories") are too under-specified in terms of ontological content (assumptions about system, structure, agency, and relationships between actors) to be used to identify the particular causal factors at work, or reason the specific mechanisms involved in the explanation of empirical relationships. As became evident in Chapters 7–10, this is for subsequent middle-range theories to generate.

The still crucial epistemological contributions of FPA to the study of foreign policy-making, decisions on caveats included, is rather to provide direction, cohesion, inspiration, and bottom-up insights to the study of political coalition-behavior. This *directional contribution* of the FPA approach comes in three packages: First, FPA directs attention to and conceptualizes policy-making processes often neglected in comparative analyses of covariation related to caveats, and in research emphasizing burden-sharing negotiations in alliance politics. By directing attention to states' coalition-behavior in terms of (i) identifying, constructing, and widening the SPM, (ii) prioritizing objectives and goals, (iii) composing packages of policy instruments to serve preferences, and (iv) actual implementation behavior, FPA invites us to study actual decision-making behavior and thus to discover the institutional, organizational, social, and psychological mechanisms at work when decisions are made on coalition participation and the application of caveats.

Second, the approach of FPA is extremely ambitious in urging the integration of several levels of analyses in the search for causal input-factors to the decision-making process. In the explanation of decision-making outputs, FPA directs us to include the theoretical input that may explain patterns of caveats from the levels of global politics, domestic politics as well as from the institutional and individual level of decision-making and implementation. Indeed, in the FPA approach to the politics of caveats, we are invited to study how domestic and global politics interplay in influencing the foreign policy decision-making processes that produce decisions on the participation in coalition forces. We are further encouraged to research how attributes of the state governmental apparatus and key decision-makers influence perceptions of SPM and preferences, and the choice of policy instruments, caveats included. In an integrated FPA model, all these levels of analyses are theorized for the promise of discovering how global and domestic causal impulses interact through decision-making and implementing processes (Fermann 2013: 117–128). This is not to say that we are to include all these levels of analyses and phases of decision-making and implementation in each particular research project. The holistic ambition of the FPA approach must rather be understood as a collective and cumulative research responsibility.

Finally, in addition to insisting on a multi-level and decision-making approach to the explanation of foreign policy outputs, the FPA approach provides directional advice on how to *select and translate bodies of theory* from the Political Science branches of International Relations, Comparative Politics, Political Behavior, and Public Administration and beyond for the purpose of studying foreign policy decision-making and the politics of caveats. Critical selection is necessary due to the vast range of bodies of literature available in Political Science, Political Sociology, Political Psychology, organizational theory, and beyond (Gerner 1992). Theory-translation is often necessary because theorizing in other branches of Political Science are related to adjacent or completely different political phenomena such as international cooperation and conflict, elections and public opinion, and areas of public policies other than foreign policy. Hence, the FPA approach does even to pay lip service to concerns about research parsimony, but is due to its ontological realism nevertheless likely to contribute to the filling of some crevasses in the present knowledge on the politics of caveats.

In FPA, we have thus to do with a multi-level, decision-making, theory-borrowing, and theory-translating approach to the study of foreign policy-making and state-behavior in alliance politics and coalition operations. This makes FPA an eclectic and theory-pragmatic approach, and provide us with considerable latitude to adapt the choice and translation of theories to the phenomenon under study and particular research questions.

The *second step* in the elaboration of the framework for analysis was to act on the advice of the FPA approach and review and select *bodies of ontologically rich theory* from multiple levels of analyses. Epistemologically, the contribution of fully fledged theories ("auxiliary hypotheses," in Lakatosian language) to the analytical framework is to *identify promising explanatory variables and explain the mechanisms at work* between the explanatory variables and the phenomenon under scrutiny. To protect the "rear-guard" of the deductive bridge in the analytical framework, the selected theories are required to harmonize with the ontological tenets of the FPA approach (the "hard core"), either directly or through appropriate ontological translation. To prepare for the subsequent hypotheses-generating move, we should attempt to clarify by way of theoretical reasoning how some factor "X" and some mechanism "m" are likely to condition and mediate certain coalition and caveats-behavior ("Y").

In Chapters 8–10, the FPA approach was supplemented with several carefully selected bodies of theories with the purpose of shedding light on exactly how caveats may be used as a political instrument to secure the political compromises necessary to make participation in coalition forces feasible. At the level of *global politics*, we discussed the literature as to how alliance dynamics and security dilemmas could help explain coalition participation and caveats-behavior in terms of the balancing of divergent interests in alliance politics. At the level of *domestic and governmental politics*, we applied theory on intra-governmental relations, institutions and decision-making processes to explain how structural conditions and mechanisms of risk aversion may influence the application of caveats as an instrument for the brokering of domestic winning political coalitions for the decision to participate in coalition forces. Finally, concerning the domain of the *politics of implementation*, we discussed how principal-agent literature on civil–military relations might explain caveats in terms of national political control of military implementation.

Recall that the primary implication of approaching national reservations on the use of force as a foreign policy instrument is that caveats are framed as a tool to make possible participation in coalition forces, and to optimize the national military footprint in the coalition within the political constraints prevailing. Without the policy-option of caveats, more states would likely reserve themselves against any serious military participation or withdraw troops prematurely. In this regard, the FPA approach to the politics of caveats is in considerable contrast to the lion-share of the scholarly literature and the predominant political and military discourse, which tend to approach caveats as a burden-sharing problem, a drain on military efficiency, and a threat to the implementation of the political mandate of the military coalition. While the IR approach to international negotiations and burden-sharing sees caveats mainly as part of the problem, the bottom-up and multi-level approach of FPA opens up for less dystopic implications. Indeed, an overriding conclusion from the study is that the foreign policy instrument of caveats may be less of a liability for the military execution of the political mandate of the coalition than it is a vital means for recruiting allies to the coalition in the first place. While the use of restrictive caveats does signal reluctant participation, caveats may allow hesitant and lukewarm states to participate in coalitions they otherwise would have chosen to abstain from. Such framed, inconvenient caveats may be a blessing in disguise for multinational military operations—if fighting alone is not an option. This key suggestion, in turn, is a reminder that facts about particular phenomena do not interpret themselves but is assigned meaning depending on the interpretive framework applied.

The *third and final contribution* to the framework for analysis of the political study of caveats was the deduction of several empirical propositions (hypotheses) through several lines of theoretical reasoning on three levels of analyses (Chapters 8–10). The theoretical lines of reasoning led to three, 23 and 14 hypotheses on the levels of "alliance politics," "domestic and governmental politics," and the "politics of implementation," respectively. We offered the empirical propositions for further theoretical consideration and empirical research, as assembled below (Table 12.1).

The three-tier framework for analysis summarized above was not suggested as a comprehensive blueprint for research on the politics of caveats, but instead as a set of directions on how to reason systematically on the politics of caveats. Having said that, the decision to frame caveats as a foreign policy instrument, the emphasizing of the study of actual decision-making and implementing processes, and the theorizing of how these processes are impacted by attributes of the global, domestic, institutional, and the individual decision-making levels of analyses (FPA) are fundamental to the analytical framework. Hence, the directions elaborated, will influence subsequent design decisions.

Still, in the subsequent effort to select, translate, and reason particular theories on each one of several levels of analyses, the number

Table 12.1 Caveats and coalition participation—empirical propositions for further consideration

Empirical propositions deduced[a]

"Alliance Politics" hypotheses (AP-H1–3):

AP-H1: *"Abandonment" hypothesis:* States primarily concerned with the risk of abandonment are inclined to provide unconditional military support to the coalition (no caveats)

AP-H2: *"Entrapment" hypothesis:* States primarily concerned with the risk of entrapment are not inclined to provide military support to the coalition (no participation whatsoever, caveats irrelevant)[b]

AP-H3: *"Optimal trade-off" hypothesis:* States primarily concerned with the balancing of the dual risks of abandonment and entrapment are inclined to apply caveats to an extent compatible with the relative importance of the two concerns (measured application of caveats)

"Domestic and Governmental Politics" hypotheses (DGP-H1–23):

DGP-H1: *"If and how" hypothesis:* When assessing invitations to participate in coalition forces, states are inclined to let external considerations related to threats to security and alliance politics decide *whether* to contribute, and domestic concerns related to political feasibility decide *how* to contribute[b]

DGP-H2: *"Dual-purpose" hypothesis:* States apply caveats not only to balance security dilemmas in alliance politics but even more to reconcile domestic political disagreements

DGP-H3: *"Latitude" hypothesis:* Decision-makers with a high degree of autonomy in relation to other foreign policy decision-making institutions, and with considerable discretion to make decisions on coalition participation are inclined to contribute substantial military forces to the coalition[b]

DGP-H4: *"Veto, no impact" hypothesis:* Parliamentary veto-powers on decisions related to use of force abroad does not make a difference as to *whether* and *how* states participate in coalition forces[b]

DGP-H5: *"Reservations in numbers" hypothesis:* The more actors (veto-players) having access to the domestic decision-making process, the more demanding it is to reach a compromise solution on coalition participation and thus a greater inclination to apply caveats on the contribution as part of a political compromise

DGP-H6: *"Deep compromise" hypothesis:* In democratic coalition governments in multi-party parliamentary systems, the challenge of having military coalition participation approved by the other parties in the government coalition is amplified and the inclination to apply restrictive caveats even stronger to protect the contribution to the coalition forces

DGP-H7: *"Ideological polarization" hypothesis:* The larger the ideological distance between the veto-players in the decision-making process, the more demanding it will become to work out a compromise solution. The prospect of deadlock increases the probability that caveats be attached to force contribution

DGP-H8: *"Legitimate cause" hypothesis I:* A Center-Left coalition government is inclined to turn down an invitation to contribute militarily to a narrowly composed "coalition-of-the-willing" coalition, which rests on a weak foundation in International Law, and fails to find support in a credible humanitarian narrative[b]

(continued)

Table 12.1   (continued)

Empirical propositions deduced[a]

DGP-H9: *"Legitimate cause" hypothesis II*: A Center-Right coalition government is inclined to make an unreserved military contribution to a coalition force if considerable economic and security interests are at stake

DGP-H10: *"Legitimate cause" hypothesis III*: For a Center-Left government to accept military participation in a coalition on mainly economic and security grounds, considerable concessions are likely to be made in the form of restrictive reservations on the use of force

DGP-H11: *"Legitimate cause" hypothesis IV*: For a Center-Left government to accept military participation in a coalition on mainly economic and security grounds, the military contribution would likely be reduced to a token one, depending on what other concerns need to be taken into consideration[b]

DGP-H12: *"Cosmopolitan outlook" hypothesis I*: The inclination and primary position of a coalition government party with a cosmopolitan/globalist outlook is to accept full participation in coalition force without any strings attached (no caveats)

DGP-H13: *"Cosmopolitan outlook" hypothesis II*: The subsidiary position of a coalition government party with a cosmopolitan/globalist outlook would be to prefer the application of caveats on the military contribution to no participation at all

DGP-H14: *"Communitarian outlook" hypothesis I*: The inclination and primary position of a coalition government party with a predominantly communitarian/nationalist outlook is to decline any participation in coalition force[b]

DGP-H15: *"Communitarian outlook" hypothesis II*: The subsidiary position of a coalition government party with a communitarian/nationalist outlook is to accept some degree of military participation provided that the government coalition partners accept considerable restrictive caveats

DGP-H16: *"No unity, no leverage" hypothesis*: A party split by disagreement on war participation and thus incapable of blocking or conclusively influence its coalition partners, is unlikely to gain acceptance for the application of caveats

DGP-H17: *"Unlikely case" hypothesis*: A small party in a coalition government, with a communitarian/nationalist outlook on foreign affairs and in control of the ministry of foreign affairs, is inclined to favor a smaller military contribution to the coalition and to attach a degree of restrictive caveats to it

DGP-H18: *"Institutional prerogative" hypothesis*: The inclination to apply caveats on force contributions is stronger if the decision-making power rests in the ministry of foreign affairs rather than in the ministry of defense

DGP-H19: *"Institutional connect" hypothesis*: The inclination to apply caveats on force contributions is weaker if the foreign policy-making process is close or strongly integrated into the multilateral politics of security organizations

DGP-H20: *"Variable-sum" hypothesis*: The more divisible and structurally open for compromise the issue of coalition participation is, the more inclined is the coalition government to apply caveats on the military contribution

DGP-H21: *"Zero-sum" hypothesis*: The more indivisible and structurally closed for compromise the contentious issue of coalition participation is, the less inclined is the coalition government to agree upon a military contribution[b]

(continued)

Table 12.1 (continued)

Empirical propositions deduced[a]

DGP-H22: *"Game changer" hypothesis:* The more skilled the political entrepreneur (mediator) is in framing the political instrument of caveats as a tool capable of transforming the contentious issue of participating in coalition force from one of zero-sum (indivisible) to one of variable-sum (divisible), the more likely it is that restrictive caveats are applied to secure coalition participation

DGP-H23: *"Facing political extinction" hypothesis:* Where the future of the government is at stake on the question of coalition participation, foreign policy decision-makers will either back down from participating in the coalition operation or—depending on circumstances—offer a minimal force or a force with considerable restrictions on the use of force

"Politics of Implementation" hypotheses (PI-H1–14):

PI-H1:*"Principal's dilemma" hypothesis:* Political efforts to maximize military effectiveness using restrictive caveats, is likely to reduce military efficiency

PI-H2: *"No-arm-tied-on-the-back" hypothesis I: To the extent the military agent act as veto-player in the planning and decision-making on the participation in the coalition, this decreases the political principal's inclination to apply restrictive caveats on the national contingent*

PI-H3: *"No-arm-tied-on-the-back" hypothesis II: To the extent, the political principal applies restrictive caveats to national contingent the military agent is inclined to apply a flexible interpretation of the caveats imposed*

PI-H4:*"Too strong for comfort" hypothesis: The more robust the coalition force, the more inclined are political principals to apply restrictive caveats on the national contingent*

PI-H5:*"Too close for comfort" hypothesis: The more integrated into other nations' forces the national contingent becomes, the more inclined are political principals to apply restrictive caveats*

PI-H6: *"Tolerance for political risk" hypothesis I: Political principals with a risk-averse mindset have a strong incentive to monitor and apply national reservations on the use of force*

PI-H7: *"Tolerance for political risk" hypothesis II: The more willing political principals are to accept the potential repercussions from domestic politics, the less inclined they are to apply restrictive caveats*

PI-H8: *"Tolerance for political risk" hypothesis III: The more willing political principals are to accept the risk of being singled out for criticism by allies, the more inclined they are to offer a limited military contribution or apply restrictive caveats*

PI-H9: *"Prospects for gain and risk acceptance" hypothesis: The more the political principal frames the participation in the coalition concerning prospect for gains, the less risk the principal is willing to accept, and thus more inclined to apply restrictive caveats*

PI-H10: *"Prospects for loss and risk acceptance" hypothesis: The more the political principal frames the participation in the coalition force in terms of prospect for loss, the more risk the principal is willing to accept, and thus less inclined to apply restrictive caveats*

PI-H11: *"Police-patrol oversight" hypothesis: Political principals preoccupied with avoiding mission creep and goal-slippage are inclined to apply restrictive caveats on the military contingent using the assertive mechanism of "positive command."*

(continued)

Table 12.1   (continued)

Empirical propositions deduced[a]

PI-H12: "Fire-alarm oversight" hypothesis: Political principals preoccupied with contingent's operational flexibility and efficiency are inclined to delegate authority to the military to make decisions on the use of force through the mechanism of "command by negation," to the extent applying restrictive caveats at all

PI-H13: "Crisis-management" hypothesis: Political principals accepting high levels of risk due to prospects for huge losses are inclined to remove restrictive caveats in the chain of command, apply permissive caveats on the use of force, and/or strengthen the military contribution with more troops

PI-H14: "Political insurance" hypothesis: Political principals in need of establishing "plausible deniability" in case the mission backfires are inclined to delegate full authority to military agents to decide how to execute the military operation

[a]Eight out of the 40 hypotheses formulated do not relate to caveats in any direct sense, but to the initial government-decisions as to *whether* and with *what forces* to participate in coalition force. These propositions are marked "b" in the table and include AP-H2, DGP-H1, DGP-H3–4, DGP-H8, DGP-H11, DGP-H14 and DGP-H21. The remaining 32 hypotheses relate to the main task of the study, the conditions for governments' application of caveats on their coalition force-contributions. Recall that the issue of caveats is limited to governments' decisions on applying national reservations on the use of force on contingents in the field.

of design decisions multiply and becomes contingent on the particular research-purpose, as do the number of potential causal inferences and operational choices as to how the independent variables be measured before any gathering of empirical data. In dealing with the triple challenge of being theoretically comprehensive in scope, deductively meticulous and operationally precise, we have made several compromises.

Starting with the last step before engaging in empirical research, note that the independent variables in the 40 hypotheses deduced are not all nearly enough specified to represent a precise guide to empirical measurement. The independent variables range from the very precise operationalization to the abstract conception. For instance, in hypotheses, AP-H1 and AP-H2, the independent variables "fear of abandonment" and "fear of entrapment" need further specification before any empirical research. Further specification is also required for the independent variable of "plausible deniability" in hypothesis PI-H14, and "cosmopolitan/globalist outlook" in DGP-H13. Until such vague independent variables are properly operationalized, they cannot be used for empirical research.

The relevant associated empirical propositions should thus for the time being be considered macro hypotheses requiring further specification.

Notable is also the fact that we have not phrased the 40 hypotheses according to some singular standard of formulation. In an applied empirical study, such lack of consistency is problematic. Less so, if at all, in a programmatic study such as ours, which offers hypotheses as *raw material* to the research community on the expectation that researchers will need to adapt hypotheses and deductive lines of reasoning to the requirements of their particular research. It is impossible to outguess what formulation of hypotheses might serve particular research projects the best. Arguably, the heterogeneity of the several hypotheses-formulations is also a reflection of the present state of the research literature in the field. Regardless, the semantic heterogeneity of hypotheses-formulations is an indication of the suggestive nature of the empirical propositions.

Less of a limitation and perhaps more of a useful addition is the fact that 8 out of 40 hypotheses do not deal with conditions for national reservations on the use of force in the field, but with the preceding government-decisions as to *whether and what to contribute to the coalition.* In the reviewing of the caveats-literature and several bodies of theory relevant for more generic purposes, the explanatory ideas on the two former government-decisions and related phenomena came up. Sometimes as a separate issue, at other times intertwined with the subsequent issue of caveats. Suspecting that such inferences might be useful for some research purposes, we decided to allow this deviation from the central dependent variable of the study to reflect in the deduction of hypotheses. Empirical propositions dealing not directly with conditions influencing states propensity to apply caveats include AP-H2, DGP-H1, DGP-H3-4, DGP-H8, DGP-H11, DGP-H14, and DGP-H21. The remaining 32 hypotheses relates to the main task of the study, the conditions for governments' application of national reservations on their contingents' use of force in coalition warfare.

More serious a limitation is the varying extent to which the full range of hypotheses formulated enjoys the support of a strong theoretical argument, which reasons precisely how a condition "X" causally impacts another phenomenon "Y." For instance, in hypotheses AP-H1-3 the mechanisms involved in the security dilemma of alliance politics were argued in considerable detail, as were the prospects for loss and gain mechanisms linking independent and dependent variables in hypotheses PI-H6-10. However, in some other instances, the deductive link between theory and hypotheses is considerably weaker. For instance, in several of the input hypotheses discussed at the level of domestic politics, additional theory is required to explain properly the mechanisms at work between the independent and dependent variables. The reason such hypotheses still have been included is the multiple indications in the literature on foreign policy-making that certain factors may affect caveats-behavior, however, without offering a fully developed argument as to how this is going about.

Deficiencies in explaining the mechanisms at work in a hypothetical relationship is an invitation to introduce additional theory and literature to fill the gaps in the deductive reasoning prior to any empirical testing. Additional literature is also required where increasingly important developments (such as the possibly growing cleavage between globalist and communitarian ideological outlooks in foreign policy) are not properly accounted for in caveats-relevant literature. What in particular is required is the tapping into theory capable of explaining the possible causal links between external and domestic conditions to the foreign policy-making process. We need to theorize more thoroughly on how global and domestic input-factors are institutionally received, politically perceived, made sense of, and acted upon in the foreign policy-making processes.

In this, Graham T. Allison's classical work on bureaucratic politics still has much to offer, as a point of departure, to the understanding of how the institutional structure and division of labor of foreign policy-making processes shape the external input to the foreign policy-making process (Allison and Zelikow 1999). If the decision-making output is a negotiated result of infighting between governmental institutions with different mandates and kinds of resources ("bureaucratic/governmental politics"), it is crucial to understand how the organizational make-up and the rules regulating the foreign policy-making process affect decisions on coalition participation.

Crucial mechanisms linking causal input-factors to foreign policy output on caveats may also be found in the standard operating procedures (SOP) all complex organizations use to cope with excessive demands on decision-making capacity. In the organizational-process model, Allison theorizes how policy-response and implementation of policies are directed by a programmed procedure as soon as it is determined what problem-solving SOP fit the purpose (Allison and Zelikow 1999). This line of reasoning can be useful to unravel mechanisms between input-factors to the foreign policy-making process and decisions on goals and means, but also to better understand the mechanisms at work in the actual implementation of caveats-policies in the field. The caveats applied may be due to some strategic culture or military doctrine that is institutionalized in particular operational practices concerning the use of force (Ingesson 2017; Fermann and Inderberg 2013). This is a separate and complementary line of reasoning to the principal-agent approach that inspired the deduction of several hypotheses in the chapter on the politics of implementation (Frost-Nielsen 2016; Auerswald and Saideman 2014).

Furthermore, the distinction made in the FPA approach between perceptions of SPM, the choice of preferences and goal-serving policy instruments, and the actual implementation of policies offers ample additional opportunities to apply insights from Social Psychology (Ross et al. 2010) and Political Psychology (Levy 2003). "Just" as Jack S. Levy's prospects theory (1997, 1996) was used to explain the connection between risk averseness and the application of caveats in foreign policy, we may apply theories on confirmation bias, cognitive attributes, group-think, and decision-makers' prior experience and career to explain the causal mechanisms at work between external input-factors and decision-making output in the foreign politics of caveats.

Hence, in a causal chain originating in input from global politics and domestic politics, and finally ending in the actual practicing of caveat policies in the field, the several intermediate stages in the foreign policy-making and implementing processes is a fruitful place to seek answers to the important "how"-question in causal explanation (Thomas 2011: 513). As for the as crucial "what" and "why"-questions, they are answered by means of the reasoned identification of some independent variable, and through the establishment of an empirical relationship between a particular independent variable and the phenomenon of caveats. The question remains, how do phenomenon-, approach-, and field-specific attributes of the present empirical research program impact our preferred choice of strategies for the gathering and analyses of data in the context of the politics of caveats?

METHODOLOGICAL CONTRIBUTIONS: DESIGNING EMPIRICAL RESEARCH

The *final main contribution* of the research program relates to the cluster of methodological issues that need to be addressed in connection with the gathering and analyses of data. At the outset, we suggested that solutions to methodological challenges must take into account the key characteristics of the policy domain we are studying (security policy, foreign policy), the maturity of our field of research (politics of caveats), and the analytical approach applied (FPA).

As to the *gathering of data*, we identified two main challenges. One is the lack of a common conception of caveats across studies, as discussed above. Without a common yardstick as to how to measure national reservations on the use of force, we will not be able to build a systematic database on the phenomenon. The lack of agreement on the core concept of caveats tells us that the study of the politics of caveats still is in a nascent stage of development. Our contribution to building systematic and comparable data on caveats was to reason a three-dimensional conception of caveats that distinguished caveats from adjacent phenomena, and made it possible to differentiate between different kinds of caveats.

The second impediment to the gathering of data is that not all relevant information on caveats are readily available. In particular, this may include nation- and coalition-specific information on RoE, command-structures, and the specific content of multilateral settlements on burden sharing in NATO—so essential for our operational definition of caveats. We argued that in a policy-area pervaded by concerns about national security and political accountability, we should expect some information to be classified or hard to find. We should also expect to see variations in how willing states are to offer information on coalition participation and in particular on national RoE. We have also experienced that information which is hard to retrieve from national authorities, may be available from coalition sources. Nevertheless, political and military sensitivity provides an incentive to control information flows and restrict access to some caveats-relevant information.

The challenge of gathering solid data multiply when we put the approach of FPA to action. This is because FPA invites the study of several levels of analysis and multiple phases in decision and implementing processes. This may include the attempt to map what considerations influenced the assessment of SPM, the deciding on goals and preferences, the calibration of policy instruments, and the structuring of the interface between political decision-making and military implementation (Fermann 2013: 89–139). This kind of information is rarely available in some pre-made generic database, but in high demand among FPA researchers conducting a process tracing analysis of policy decisions and implementation behavior (Bennett and Checkel 2015). The more crucial it is to look at ways to work around obstacles to data gathering in the particular field of coalition participation. What can we do to circumvent political and military incentives to protect caveats-relevant information against exposure?

Aside from being equipped with a precise conception of caveats and use available open sources and material, we have argued that caveats-research benefit from including political scientists with a military background in the research team and establish research cooperation with military schools. Well-placed researchers and military academics can get access to data in the shape of unclassified documents, through interviews and even participatory observation. The officer-turned-scholar is in a position to utilize military and institutional "local knowledge" and professional skills to find the richest sources of information, get access to key military personnel, ask the door-opening questions, and to better interpret, contextualize, and follow-up on the feedback they receive from their interviewees.

By getting access to the higher echelons of military command, the military academic may also retrieve crucial information on what is going on at the interface between policymakers and top military brass (Auerswald and Saideman 2014; Henriksen 2007). Such information gathering may reveal the existence of informal caveats of the restrictive or permissive kind. Information gathered from mid- and lower-level officers may reveal the existence of national RoE deviating from coalition RoE, the behavior of national red card-holders in the chain of military command, and particular conditions on the use of force codified in NATO burden-sharing settlements or bilateral force agreements.

Caveats-relevant data on decision-making and implementing processes is unlikely to all come in the form of hard evidence, the proverbial "smoking gun." Circumstantial evidence provided through unconfirmed statements, rumors, and perspective-dependent sources is a legitimate empirical part of the picture we form about hard-to-establish framework factors; this or that decision-maker; certain kind of priorities; and particular assessments of means-end relationships made behind closed doors. While having a sharp eye on the reliability of the sources and the validity of the information gathered, we make use of all kinds of retrievable data to the extent it can be justified by the research questions posed, the approach applied, and the operational definitions decided on. Along with theoretical insights, we systematically triangulate hard and circumstantial evidence to establish the relevant facts, and subsequently, to argue causal relationships.

Concerning the *empirical analyses of data*, we have been limited in our methodological choices by the scattered availability of data for generic inquiry and influenced by the decision-making approach of FPA, which is at the core of our empirical research program. Both considerations pointed in the direction of single and comparable-cases research designs, including Qualitative Comparative Analysis (QCA). Considered collectively, the various case designs are capable of;

1. making theoretical generalizations by providing analytical answers to the question as to what some case is a case of (Moses and Knutsen 2012: 143);
2. adjusting the boundaries of the scope conditions for the explanatory theory in question (Harris 1997);
3. establishing control for third variables by means of a pattern-matching logic of causal inference (Lijphart 1975: 164);
4. dealing with an outcome which can be explained in terms of different sets of combinations of causes ("multiple conjectural causations," INUS conditions) (Ragin 1987; Mackie 1965);
5. coping with several levels of measurement (nominal, ordinal, interval) (Rihoux and Ragin 2009: 33–122); and of
6. tracing causal paths between independent and dependent variables to validate theoretical assumptions on mediating mechanisms (Checkel 2006).

The main methodological contribution of the study is the Chapter 11 *review of the case-design literature* as regards theory development and theory testing. The account was scattered with foreign policy and caveats-relevant illustrations to indicate applicability. For *theory-building purposes*, we first discussed how the *deviant case* design be used to refine existing theory to account for anomalies, either by broadening or limiting the scope of the theory. Furthermore, we suggested the inductive version of the *plausibility probing* design be applied as a pilot study to assess "the details of a particular case in order to shed light on a broader theoretical argument […] allowing the researcher to sharpen a hypothesis or theory, or to refine the operationalization or measurement of key variables" (Levy 2008: 6). For theory-developing purposes, the inductive version of the *process tracing* design direct us to "practice trace" real-life decision-making processes (Pouliot 2015) to "capture causal mechanisms in action" (Bennett and Checkel 2015: 9), to better understand how certain inputs are associated with certain outcomes.

Common to the three case designs as presented, is the generalizing move of procedures. Plausibility probing for inductive purposes is not limited to a particular generalizing focus as it, in principle, covers any inductive reasoning from case-specific facts used to develop precise conceptual, typological and ontological constructs for the subsequent study of a class of cases. Process tracing as "practice tracing," on the other hand, is more specific in its theory-building purpose. The case study procedure of "practice tracing" borrows from historicity, "the temporal structuration of social actions and processes" (Hall 2007: 82), in the attempt to conceptualize and theorize the social processes that causally connects independent and dependent variables.

By theorizing the timeline and causal path of case-particular decision-making processes, the "practice-tracing" researcher makes sense of the connecting and mediating mechanisms at work, without which we cannot fully understand the presence of cross-unit empirical patterns. Recall that understanding the causal chain of decision-making and implementing processes is at the core of the FPA approach. As crucial, in the still-nascent study of the politics of caveats, theorizing and testing of hypotheses on the mechanisms that may, or may not constitute the causal path from "X" to "Y" is an insurance against spurious inference in comparable-cases designs, and in statistical large-N studies further down the road.

Turning to the *theory-testing* capabilities of single and comparable-cases designs, we emphasized that in single and comparable-cases designs (except process tracing), causal inference is assessed through the calculated, biased selection of cases based on the theoretical argument under consideration. In the test of a theory, it is essential to decide whether the case up for testing qualifies as a "fitting" empirical plausibility probe or some considerably more demanding crucial case. In case studies, causal inference relies on us selecting a particular case fitting both the argument and the case design well. Recall that the aims of theory-testing procedures are to "strengthen or reduce support for a theory, narrow or extend the scope conditions of a theory, or determine which of two or more theories best explains a case, type or general phenomenon" (George and Bennett 2005: 109).

Applied for theory-testing purposes, in the *plausibility probing* design we select a novel but undemanding "theory-confirming" case (Lijphart 1971: 691) to "investigate the degree to which a given case fits a general proposition" (Moses and Knutsen 2012: 137). This particular case selection criterion induces us to look for a case that is likely to illustrate our theoretical arguments and initially confirm our hypotheses, thus reassuring us that we are heading in the right direction (Eckstein 1975). Such assurance may, in turn, justify more time-consuming investments in theory-infirming (most- and least-likely case study), deviant, process tracing, and comparable-cases designs (Levy 2008: 6).

Theory-testing *crucial case* design is the most demanding single-case test (Eckstein 1975). The selection procedure mimics the well-constructed, decisive experiment by selecting a well-chosen case, which can "provide strong support for, or falsify, a given theory" (Moses and Knutsen 2012: 139). However, recall that for all practical purposes cases that provide "the most definitive type of evidence" for the falsification or validation of a theoretical argument, are hard to find in the social and political world (George and Bennett 2005: 120). As a still demanding but more realistic alternative, Eckstein suggests studying *most-likely* and *least-likely* cases. In a most-likely test, the independent variables posited by a theory are at values that strongly posit an outcome or posit an extreme outcome, while in a least-likely case, the independent variables in a theory are the values that only weakly predict an outcome or predict a low-magnitude outcome. The strongest possible supporting evidence for a theory in a tough test set-up as mentioned is "a case that is least-likely for that theory but most likely for all alternative theories, and one where the alternative theories collectively predict an outcome very different from that of the least-likely theory" (George and Bennett 2005: 121). In the least-likely case design, the assumption and the inferential logic is that "if a theoretical argument can make it there it can make it anywhere." In the most-likely case design, the inferential logic is rather that "if the theory tested cannot make it there, it cannot make it anywhere" (Levy 2008: 12).

Also in *comparable-cases design*, we achieve control and make causal inference by means of case selection. While statistical control for third variables relies on the random selection of a representative sample of cases and probabilistic relationships, comparable-cases analyses rely on a pattern-matching logic of causal inference that requires the careful selection of cases so as to "maximize the variance of the dependent variables and to minimize the variance of the control variables" (Lijphart 1975: 164). Recall that the main pitfalls of comparable-cases analysis are the sampling bias in the selection of cases resulting from relaxing of comparability criteria, and the over-determination problem arising when the comparative analysis depends on too few cases relative to the number of explanatory variables.

Comparable-cases research designs come in the basic shapes of the Method of Difference (MD), the Method of Agreement (MA), the Joint

Method of Agreement and Difference (JMAD), and the Method of Concomitant Variation (MCV) that were reasoned by John Stuart Mill (2002 [1891]), and to varying degrees iterated and elaborated on in the "different systems/similar systems" approaches (Przeworski and Teune 1970); the "comparable-cases strategies" (Lijphart 1975); the "focused comparisons" (Hague et al. 1998: 280); the case-oriented comparisons (Ragin 1987); and the "method of systematic comparative illustration" (Smelser 1973). Comparable-cases designs thus come in different configurations but unite in the ambition to widen the empirical basis for making causal inferences while simultaneously preserving some of the contextual richness typical of single and few cases designs. The latter is in part what makes comparative case study not merely an intermediary step toward the statistical large-N study, but a design with its particular epistemological advantages.

In Chapter 11, we discussed *John Stuart Mill's four basic approaches to comparative control for third variables* in some detail. For several reasons, we concluded that Mill's comparative matching strategies are better suited for *eliminating* explanatory candidates than with confidence confirming causal relationships (Moses and Knutsen 2012: 111). Hence, Mill's methods are probative, and not definitive as regards theory testing. At this point, we found ourselves at a methodological crossroads. If we have had a large enough number of units/cases at our disposal, we could have exploited statistical techniques for control of third variables in research on caveats. Since our field of study does not yet possess systematic, valid, and reliable data to the extent necessary to run statistical analyses we were left looking for other alternatives.

Reviewing the comparable-cases literature, we came to conclude that *Qualitative Comparative Analysis* (QCA) is a promising design for our theory-testing research purposes. Preferably in combination with single-case designs. QCA excels in the 5 to 75 cases range and fits well with the data situation in our particular field of research. QCA can deal with research questions engaging too many cases (double-digit) for researchers to keep all the case knowledge in their heads, but still too few cases for conventional statistical techniques. QCA requires familiarity with cases and some in-depth knowledge. Simultaneously, QCA is capable of detecting decisive cross-case patterns, the established domain of quantitative analysis. Furthermore, QCA can handle data on the nominal, ordinal, and interval level of measurement. This corresponds to crisp-set, multi value-set, and fuzzy-set QCA (Ragin 2000). While the technicalities differ, the basic procedure of the three QCA techniques is quite similar and explained in the previous Chapter.

QCA is equipped to deal with complex causation in what Charles C. Ragin terms "multiple conjectural causation" (1987). We may apply the pattern-matching strategies of QCA to uncover "several different sets of causal conditions that may lead to the same outcome" (George and Bennett 2005: 162–163). In other words, to uncover "different configurations of factors conditioning patterns of [caveats] through different causal paths" (Platt 2007: 115). Arguably, explanatory complexity is the rule rather than the exception in the study of social and political phenomena; foreign policy decision-making included (Haney 2002). The multi-level and decision-making approach of FPA encourage us to include and prepare for such causal complexity in the subsequent choice of several bodies of theory and research methods. Recall also that research on the foreign politics of caveats currently lacks the data required to run statistical analyses of caveats-relevant relationships. For such reasons, we concluded that QCA and a research program grounded in the FPA approach would be an extremely good fit.

Our final contribution to the methodology discussion was to argue that *QCA is combined with the single-case design of process tracing* to establish an even stronger theory-testing strategy. The main advantages of QCA are the capacity to deal with many cases and uncover different sets of conditions (INUS conditions) explaining the same outcome. What is lacking in QCA, however, is the capacity to trace causal *processes*. This is where the process tracing single-case design comes to rescue in its unique capacity to "capture causal mechanisms in action" (Bennett and Checkel 2015: 9). Process tracing on the level of the single case allows us to test whether the theoretical mechanisms assumed or argued are part of the causal chain between two phenomena as proposed in the hypothesis. This is crucial because we otherwise cannot be reasonably sure that covariation equals causation.

In process tracing, we understand causation as a causal path made up of a shorter or longer string of mechanisms connecting some independent and dependent variable. More specifically, we may define process tracing as "the analysis of evidence on the process, sequences, and conjectures of events within a case for either developing or testing hypotheses" (Bennett and Checkel 2015: 7). By adding process tracing to our research tool-box, we engage in the uncovering of "linear causal processes embedded in time" (Moses and Knutsen 2012: 134), where

"A causes B, B then causes C, C then causes D and so on" (Checkel 2006: 363). Thus, process tracing shifts our epistemological focus from "what happened, to how and why it happened" (Moses and Knutsen 2012: 225). In particular, the process tracing case design has a comparative advantage in the empirical analysis of "decision making at the individual, small group, and organizational levels," including the analysis of leaders' perceptions, judgments, preferences, internal decision-making environment, and choices (Levy 2008: 11).

As elaborated in the previous Chapter, the process tracing case design is a door opener to the proverbial "black box" of governmental decision-making and implementing processes. As QCA responds to the multi-level explanatory ambition of FPA by allowing for "multiple conjectural causation" (Ragin 1987), we consider process tracing tailor-made to deliver on the second main tenet of the FPA approach that underpins our empirical research program. In particular, process tracing delivers by tracking the political, institutional, and psychological mechanisms at work in the translation and shaping of external and domestic input impulses to the foreign policy decision-making and implementing processes that eventually produces an outcome in terms of coalition-behavior and caveats patterns. Arguably, a best practice for theory-testing purposes would be to use QCA to get a broad overview of the causal landscape, and then apply process tracing on the level of the single case to get a high-resolution picture of significant details in the causal path researched. For theory-developing purposes, the inductive version of process tracing, practice tracing, would be an attractive first choice along with plausibility probing case designs.

The lion-share of our methodology discussion has dealt with theory-testing procedures. At the end of the road, we should remind ourselves what is up for empirical testing. It is not the FPA approach as such since we at the outset granted FPA a privileged, axiomatic, and directional status in the broader analytical framework of the research program. By constituting the epistemological "hard core" of the framework for analysis, the approach of FPA is, in principle, beyond testing. However, FPA is not beyond debate as to how fruitful or necessary the approach is for the study of the politics of caveats. What is on immediate epistemological trial though, and wide open for testing against empirical material is the multiple bodies of "auxiliary hypotheses" (middle-range theories) we may decide to use as ontological extensions of FPA and from which empirical propositions subsequently are deduced (Lakatos 1978).

REFERENCES

Allison, G., & Zelikow, P. (1999). *Essence of Decision. Explaining the Cuban Missile Crisis*. New York, NY: Longman.

Auerswald, D. P. (2004). Inward Bound: Domestic Institutions and Military Conflicts. *International Organization, 53*(3), 469–504.

Auerswald, D. P., & Saideman, S. M. (2014). *NATO in Afghanistan: Fighting Together, Fighting Alone*. Princeton, NJ: Princeton University Press.

Bennett, A., & Checkel, J. T. (Eds.). (2015). *Process Tracing. From Metaphor to Analytical Tool*. Cambridge: Cambridge University Press.

Bergen, P. L. (2011). *The Longest War*. New York, NY: Free Press.

Brophy, J., & M. Fisera. (2010). *National Caveats and Its Impact on the Army of the Czech Republic*. http://user.unob.cz/fisera/files/clanky/National_Caveats_Short_Version_version_V_29JULY.pdf.

Caldwell, B. J. (1991). The Methodology of Scientific Research Programmes: Criticisms and Conjectures. In G. Keith Shaw (Ed.), *Economics, Culture, and Education. Essays in Honor of Mark Blaug* (pp. 95–107). Aldershot: Elgar.

Checkel, J. T. (2006). Tracing Causal Mechanisms. *International Studies Review, 8*(2), 362–370.

Clark, W. K. (2001). *Waging War: Bosnia, Kosovo, and the Future of Combat*. New York, NY: Public Affairs.

Deni, J. R. (2004). The NATO Rapid Deployment Corps: Alliance Doctrine and Force Structure. *Contemporary Security Policy, 25*(3), 498–523.

Eckstein, H. (1975). Case Study and Theory in Political Science. In F. I. Greenstein & N. W. Polsby (Eds.), *Handbook of Political Science* (Vol. 7, pp. 79–137). Reading, MA: Addison-Wesley.

Fermann, G. (Ed.). (2013). *Utenrikspolitikk og norsk krisehåndtering. Oslo: Cappelen Damm Akademika*. https://www.cappelendamm.no/_utenrikspolitikk-og-norsk-kriseh%C3%A5ndtering-gunnar-fermann-9788202378691.

Fermann, G., & Inderberg, T. H. (2013). Norway and the 2005 Elektron Affair: Conflict of Competences and Competent Realpolitik. In T. G. Jakobsen (Ed.), *War. An Introduction to Theories and Research on Collective Violence* (pp. 373–402). New York, NY: Nova Science.

Findlay, T. (2002). *The Use of Force in UN Peace Operations*. Oxford: Oxford University Press.

Frost-Nielsen, P. M. (2009). *Rules of Engagement: En utenrikspolitisk case-analyse av den politiske kontrollen av norske kampfly i Operation Enduring Freedom, Afghanistan 2002–2003*. Master Thesis in Political Science, Department of Sociology and Political Science, Norwegian University of Science and Technology (NTNU), Trondheim.

Frost-Nielsen, P. M. (2011). Politisk kontroll av militær deltakelse i internasjonale operasjoner: Restriksjoner på bruk av norske kampfly i Afghanistan. *Internasjonal Politikk, 69*(3), 359–386.

Frost-Nielsen, P. M. (2013). Norske kampfly i Afghanistan 2006. In G. Fermann (Ed.), *Utenrikspolitikk og norsk krisehåndtering* (pp. 267–298). Oslo: Cappelen Damm Akademika.

Frost-Nielsen, P. M. (2016). *Betingede forpliktelser. Nasjonale reservasjoner i militære koalisjonsoperasjoner.* Ph.D. Dissertation in Political Science, Department of Sociology and Political Science, Norwegian University of Science and Technology (NTNU), Trondheim.

Frost-Nielsen, P. M. (2017). Conditional Commitments: Why States Use Caveats to Reserve Their Efforts in Military Coalition Operations. *Contemporary Security Policy, 38*(3), 371–397.

Frost-Nielsen, P. M. (2018, forthcoming). Bringing Military Conduct Out of the Shadow of Law: Towards a Holistic Understanding of Rules of Engagement (RoE). *Journal of Military Ethics 17*(1–2).

George, A., & Bennett, A. (2005). *Case Studies and Theory Development in Social Sciences.* Cambridge, MA: MIT Press.

Gerner, D. (1992). Foreign Policy Analysis. Exhilarating Eclecticism, Intriguing Enigmas. *International Studies Notes, 18*(1), 4–19.

Gerring, J. (1999). What Makes a Concept Good? A Critical Framework for Understanding Concept Formation in the Social Sciences. *Polity, 31*(3), 357–393.

Hague, R., Harrop, M., & Breslin, S. (Eds.). (1998). *Government and Politics: An Introduction.* Basingstoke: MacMillan.

Hall, J. R. (2007). Historicity and Socio-historical Research. In W. Outhwaite & S. P. Turner (Eds.), *The Sage Handbook of Social Science Methodology* (pp. 82–101). London: Sage.

Haney, P. J. (2002). *Organizing for Foreign Policy Crises: Presidents, Advisers, and the Management of Decision-making.* Michigan, MI: University of Michigan Press.

Harris, W. A. (1997). On "Scope Conditions" in Sociological Theories. *Social and Economic Studies, 46*(4), 123–127.

Henriksen, D. (2007). *NATO's Gamble. Combining Diplomacy and Airpower in the Kosovo Crisis 1998–1999.* Annapolis, MD: Naval Institute Press.

Høiback, H. (2009). The Noble Art of Constructive Ambiguity. *Oslo Files on Defence and Security, 3,* 19–39.

Ingesson, T. (2017, September 20). *Trigger-Happy, Autonomous, and Disobedient: Nordbat 2 and Mission Command in Bosnia.* The Strategy Bridge. https://thestrategybridge.org/the-bridge/2017/9/20/trigger-happy-autonomous-and-disobedient-nordbat-2-and-mission-command-in-bosnia.

Johnson, G. J. (2004). *Examining the SFOR Experience. NATO.* http://www.nato.int/docu/review/2004/Historic-Changes-Balkans/Examining-SFOR-experience/EN/index.htm.

Jones, J. L. (2004, May 7–10). *Prague to Istanbul: Ambition Versus Reality. Global Security: A Broader Concept for the 21st Century.* Center for Strategic Decision Research 21st International Workshop on Global Security—Berlin. http://csdr.org/2004book/Gen_Jones.htm.

Kay, S. (2013). No More Free-Riding: The Political Economy of Military Power and the Transatlantic Relationship. In J. H. Matlary & M. Petersson (Eds.), *NATO's European Allies—Military Capability and Political Will* (pp. 97–120). Hampshire: Palgrave Macmillan.

Koschut, S. (2014). Transatlantic Conflict Management Inside-Out: The Impact of Domestic Norms on Regional Security Practices. *Cambridge Review of International Affairs, 27*(2), 339–361.

Kreps, S. (2008). When Does the Mission Determine the Coalition? The Logic of Multilateral Intervention and the Case of Afghanistan. *Security Studies, 17*(3), 531–567.

Lakatos, I. (1978). *The Methodology of Scientific Research Program.* Cambridge, UK: Cambridge University Press.

Lijphart, A. (1971). Comparative Politics and Comparative Method. *American Political Science Review, 65*(3), 682–698.

Lijphart, A. (1975). The Comparable-Cases Strategy in Comparative Research. *Comparative Political Studies, 8*(2), 158–177.

Levy, J. S. (1996). Loss Aversion, Framing, and Bargaining: The Implications of Prospects Theory for International Conflict. *International Political Science Review, 17*(2), 179–195.

Levy, J. S. (1997). Prospect Theory, Rational Choice, and International Relations. *International Studies Quarterly, 41*(1), 87–112.

Levy, J. S. (2003). Political Psychology, and Foreign Policy. In D. O. Sears, L. Huddy, & R. Jervis (Eds.), *Oxford Handbook of Political Psychology* (pp. 253–284). New York, NY: Oxford University Press.

Levy, J. S. (2008). Case Studies: Types, Designs, and Logics of Inference. *Conflict Management and Peace Science, 25*(1), 1–18.

Lombardi, B. (2008). All Politics is Local: Germany, the Bundeswehr, and Afghanistan. *International Journal, 63*(3), 587–605.

Mackie, J. L. (1965). Causes and Conditions. *American Philosophical Quarterly, 4*(2), 245–264.

Marten, K. (2007). Statebuilding and Force: The Proper Role of Foreign Militaries. *Journal of Intervention and Statebuilding, 1*(2), 231–247.

Mello, P. A. (2014). *Democratic Participation in Armed Conflict.* Basingstoke: Palgrave Macmillan.

Meyer, T. (2013). Flipping the Switch: Combat, State-Building, and Junior Officers in Iraq and Afghanistan. *Security Studies, 22*(2), 222–258.

Mill, J. S. (2002 [1891]). *A System of Logic.* Honolulu, HI: University Press of the Pacific.

Moses, J. W., & Knutsen, T. L. (2012). *Ways of Knowing. Competing Methodologies in Social and Political Science.* London: Palgrave Macmillan.

NATO (2006, November 28). *NATO Boosts Efforts in Afghanistan.* http://www.nato.int/docu/update/2006/11-november/e1128a.htm.

Noetzel, T., & Rid, T. (2009). Germany's Options in Afghanistan. *Survival, 51*(5), 71–90.

Noetzel, T., & Schreer, B. (2009). Does a Multi-tier NATO Matter? The Atlantic Alliance and the Process of Strategic Change. *International Affairs, 85*(2), 211–226.

Platt, J. (2007). Case Study. In W. Outhwaite & S. P. Turner (Eds.), *The Sage Handbook of Social Science Methodology* (pp. 102–127). London: Sage.

Pouliot, V. (2015). Practice Tracing. In A. Bennett & J. T. Checkel (Eds.), *Process Tracing. From Metaphor to Analytical Tool* (pp. 237–259). Cambridge, UK: Cambridge University Press.

Przeworsky, A., & Teune, H. (1970). *The Logic of Comparative Social Inquiry.* New York: Wiley.

Ragin, C. C. (1987). *The Comparative Method: Moving Beyond Qualitative and Quantitative Strategies.* Oakland, CA: University of California Press.

Ragin, C. C. (2000). *Fuzzy-Set Social Science.* Chicago, IL: University of Chicago Press.

Richter, A., & Webb, N. J. (2014). Can Smart Defense Work? A Suggested Approach to Increasing Risk- and Burden-sharing Within NATO. *Defense and Security Analysis, 30*(4), 346–359.

Rihoux, B., & Ragin, C. C. (Eds.). (2009). *Configurational Comparative Methods. Qualitative Comparative Analysis and Related Techniques.* Los Angeles, CA: Sage.

Ringsmose, J. (2010). NATO Burden-Sharing Redux: Continuity and Change After the Cold War. *Contemporary Security Policy, 31*(2), 319–338.

Rosenau, J. N. (1966). Pre-theories and Theories of Foreign Policy. In R. B. Farrell (Ed.), *Approaches to Comparative and International Politics* (pp. 27–92). Evanston, IL: North-Western University Press.

Ross, L., Lepper, M., & Ward, A. (2010). History of Social Psychology: Insights, Challenges, and Contributions to Theory and Application. In S. T. Fiske, D. T. Gilbert, & G. Lindzey (Eds.), *Handbook of Social Psychology* (pp. 3–50). Hoboken, NJ: Wiley.

Ruffa, C., Dandeker, C., & Vennesson, P. (2013). Soldiers Drawn Into Politics? The Influence of Tactics in Civil–Military Relations. *Small Wars & Insurgencies, 24*(2), 322–334.

Saideman, S. M. (2018, April 3–7). *Comments to Gunnar Fermann's Paper on Coping with Caveats in Coalition Warfare: An Empirical Research Program.* Presented in the panel on The Politics of Multinational Military Operations, 2018 International Studies Association Convention, San Francisco.

Saideman, S. M., & Auerswald, D. P. (2012). Comparing Caveats. *International Studies Quarterly, 56*(1), 67–84.

Sky, E. (2007). Increasing ISAF's Impact on Stability in Afghanistan. *Defense and Security Analysis, 23*(1), 7–25.

Smelser, N. J. (1973). The Methodology in the Social Sciences. In D. P. Warwick & S. D. Osherson (Eds.), *Comparative Research Methods* (pp. 42–86). Englewood Cliffs, NJ: Prentice-Hall.

Thomas, G. (2011). A Typology of the Case Study in Social Science Following a Review of Definition, Discourse, and Structure. *Qualitative Inquiry, 17*(6), 511–521.

Trønnes, O. (2012). *Mapping and Explaining Norwegian Caveats in Afghanistan from 2001 to 2008.* Master thesis in Political Science, Department of Sociology and Political Science, Norwegian University of Science and Technology (NTNU), Trondheim.

United Nations. (2000). *Report of the Panel on United Nations Peace Operations* (The "Brahimi Report") (A/55/305-S/2000/809). New York, NY: United Nations.

van der Meulen, J., & Kawano, H. (2008). Accidental Neighbours: Japanese and Dutch Troops in Iraq. In J. Soeters & P. Manigart (Eds.), *Military Cooperation in Multinational Peace Operations: Managing Cultural Diversity and Crisis Response* (pp. 166–179). Oxon: Routledge.

Young, T.-D. (2003). The Revolution in Military Affairs and Coalition Operations: Problem Areas and Solutions. *Defense and Security Analysis, 19*(2), 111–130.

INDEX

© The Editor(s) (if applicable) and The Author(s), under exclusive
licence to Springer International Publishing AG, part of Springer
Nature 2019, G. Fermann, *Coping with Caveats in Coalition Warfare*,
https://doi.org/10.1007/978-3-319-92519-6

The manufacturer's authorised representative in the EU is Springer
Nature Customer Service Centre GmbH, Europaplatz 3, 69115 Heidelberg,
Germany. If you have any concerns regarding our products, please
contact ProductSafety@springernature.com

Printed and bound by CPI Group (UK) Ltd, Croydon, CR0 4YY
29/04/2026
02099514-0002